T0281148

# SpringerBriefs in Computer Science

SpringerBriefs present concise summaries of cutting-edge research and practical applications across a wide spectrum of fields. Featuring compact volumes of 50 to 125 pages, the series covers a range of content from professional to academic.

Typical topics might include:

- A timely report of state-of-the art analytical techniques
- A bridge between new research results, as published in journal articles, and a contextual literature review
- A snapshot of a hot or emerging topic
- An in-depth case study or clinical example
- A presentation of core concepts that students must understand in order to make independent contributions

Briefs allow authors to present their ideas and readers to absorb them with minimal time investment. Briefs will be published as part of Springer's eBook collection, with millions of users worldwide. In addition, Briefs will be available for individual print and electronic purchase. Briefs are characterized by fast, global electronic dissemination, standard publishing contracts, easy-to-use manuscript preparation and formatting guidelines, and expedited production schedules. We aim for publication 8–12 weeks after acceptance. Both solicited and unsolicited manuscripts are considered for publication in this series.

**Indexing: This series is indexed in Scopus, Ei-Compendex, and zbMATH **

Ye Yuan • Xin Luo

# Latent Factor Analysis for High-dimensional and Sparse Matrices

A particle swarm optimization-based approach

 Springer

Ye Yuan
Computer and Information Science
Southwest University
Chongqing, China

Xin Luo
Computer and Information Science
Southwest University
Chongqing, China

ISSN 2191-5768          ISSN 2191-5776   (electronic)
SpringerBriefs in Computer Science
ISBN 978-981-19-6702-3          ISBN 978-981-19-6703-0   (eBook)
https://doi.org/10.1007/978-981-19-6703-0

This Springer imprint is published by the registered company Springer Nature Singapore Pte Ltd.
The registered company address is: 152 Beach Road, #21-01/04 Gateway East, Singapore 189721, Singapore

# Preface

High-dimensional and sparse (HiDS) matrices are commonly seen in many big data-related industrial applications like electronic commerce, cloud services, social networks, and wireless sensor networks. Despite its extreme sparse, an HiDS matrix contains rich knowledge regarding desired patterns like users' potential favorites, item clusters, and topological neighbors. Hence, how to efficiently and accurately extract desired knowledge from it for various data analysis tasks becomes a highly interesting issue.

Latent factor analysis (LFA) has proven to be highly efficient for addressing an HiDS matrix owing to its high scalability and efficiency. However, an LFA model's performance relies heavily on its hyper-parameters, which should be chosen with care. The common strategy is to employ grid-search to tune these hyper-parameters. However, it requires expensive computational cost to search the whole candidate space, especially when building an LFA model on an HiDS matrix. Hence, how to implement a hyper-parameter-free LFA model becomes a significant issue.

In this book, we incorporate the principle of particle swarm optimization (PSO) into latent factor analysis, thereby achieving four effective hyper-parameter-free latent factor analysis models. The first model is a Learning rate-free LFA ($L^2FA$) model, which utilizes a standard PSO algorithm into the learning process by building a swarm of learning rates applied to the same group. The second model is a *Learning rate and Regularization coefficient-free LFA* (LRLFA) model, which build a swarm by taking the learning rate and regularization coefficient of every single LFA-based model as particles, and then apply particle swarm optimization to make them adaptation according to a predefined fitness function. The third model is a *Generalized and Adaptive LFA* (GALFA) model, which implement self-adaptation of the regularization coefficient and momentum coefficient for excellent practicability via PSO. The last model is an *Advanced Learning rate-free LFA* ($AL^2FA$) model. Before building this model, we first propose a novel position-transitional particle swarm optimization ($P^2SO$) algorithm by incorporate more dynamic information into the particle's evolution for preventing premature convergence. And then, a $P^2SO$ algorithm is utilized into the training process to make the learning rate adaptation without accuracy loss. Since hyper-parameter adaptation is done within only one full training process, thereby greatly reducing computational cost. Hence, it fits the need of real applications with high scalability and efficiency.

This is the first book about how to incorporate particle swarm optimization into latent factor analysis for implementing effective hyper-parameter adaptation. It is intended for professionals and graduate students involved in LFA on HiDS data. It is assumed that the reader has a basic knowledge of mathematics, as well as a certain background in data mining. The reader can get an overview on how to implement efficient hyper-parameter adaptation in an LFA model. We hope this monograph will be a useful reference for students, researchers, and professionals to understand the basic methodologies of hyper-parameter adaptation via PSO. The readers can immediately conduct extensive researches and experiments on the real applications HiDS data involved in this book.

Chongqing, China                                                                      Ye Yuan
June 2022                                                                              Xin Luo

# Contents

# Chapter 1
# Introduction

## 1.1 Background

With the rapid development of basic computing and storage facilities, the big-data-related industrial applications, i.e., recommendation system [1–5], cloud services [6–10], social networks [11–16], and wireless sensor networks [17–22], continue to expand. Consequently, the data involved in these applications grow exponentially. Such data commonly contains a mass of entities. With the increasing number of involved entities, it becomes impossible to observe their whole interaction mapping. Hence, a High-dimensional and sparse (HiDS) matrix [23–31, 32] is commonly adopted to describe such specific relationships among entities. For instance, the Douban matrix [33] collected by the Chinese largest online book, movie and music database includes 129,490 users and 58,541 items. However, it only contains 16,830,839 known ratings and the density is 0.22%.

Although an HiDS matrix can be extremely sparse, it possesses rich knowledge regarding desired patterns like users' potential favorites [34], item clusters [35], and topological neighbors [36]. How to precisely discover such knowledge from an HiDS matrix for various analysis tasks, i.e., accurate recommendation [1–3], community detection [37, 38], and web service selection [30, 39], is a research field with high scientific research value, economic benefits and social significance. However, conventional data analysis models [40–46] mostly fails to consider the high-dimensional, sparse and fragmented of an HiDS matrix comprehensively, thereby achieving unsatisfactory results and unavailable.

Considering existing HiDS matrix analysis models [1–3, 25–27], a latent factor analysis (LFA) model has proven to be highly efficient for addressing an HiDS matrix owing to its high scalability and efficiency. It models each row or column entity with an equally-sized latent factor (LF) vector, and then optimizes these LF vectors on an objective defined on an HiDS matrix's known data only. The well-optimized LF vectors actually embed corresponding row and column entities into the low-dimensional LF space based on an HiDS matrix's known data. Hence, it

implements precise representation to both the topological and numerical character-
istics of the concerned interaction mapping.

To date, a pyramid of LFA models has been proposed. Salakhutdinov et al. [46]
propose a probabilistic matrix factorization (MF) model, which scales linearly with
an HiDS matrix's known entry count. Yu et al. [47] propose a nonparametric MF
model with high computational efficiency on large-scale problems. Zhang et al. [48]
propose an AutoSVD++ model that generalizes the auto-encoder paradigm to LFA.
Yuan et al. [49] propose a multilayered-and-randomized latent factor model, which
adopts the principle of deep forest and extreme learning machine to enhance its
representative learning ability.

An LFA model's performance relies heavily on its hyper-parameters [50–52],
which should be chosen with care. Many optimization algorithms, i.e., a stochastic
gradient descent (SGD) algorithm [53, 54] or a single latent factor-dependent,
non-negative, multiplicative and momentum-incorporated update (SLF-NM$^2$U)
[55] algorithm, can be adopted to build an efficient LFA model. Therefore, different
LFA models correspond to different kinds of parameter. For instance, an SGD-based
LFA model depends largely on the right choice of learning rate and regularization
coefficient, or an SLF-NM$^2$U-based LFA model includes momentum coefficient and
regularization coefficient. The common practice is to employ grid-search to tune
these hyper-parameters. However, it is truly boring and expensive in both time and
computation. Moreover, since some HiDS matrices can be huge (e.g., the Yahoo!
Matrix contains 76,344,627 known entries by 200,000 users on 136,736 songs), a
corresponding LFA model costs much time to build and test during the grid search
process. It is desired to implement a hyper-parameter-free LFA model. As indicated
by prior researches [56–58], a particle swarm optimization (PSO) is an evolutionary
computation optimization due to its high efficiency and compatibility. According to
the previous work [59], it has been adopted for hyper-parameter adaptation in a
general learning algorithm. Motivated by this discovery, is it possible to incorporate
the principle of particle swarm optimization (PSO) into latent factor analysis, thereby
implementing efficient hyper-parameter adaptation for an LFA model?

To answer this question, this book proposes four hyper-parameter-free latent
factor analysis (HFLFA) models, which can implement efficient hyper-parameter
adaptation via particle swarm optimization (PSO).

## 1.2  Preliminaries

### 1.2.1  Symbol and Abbreviation Appointment

To increase the readability of this book, necessary symbols and abbreviations
adopted in this book are summarized Tables 1.1 and 1.2.

**Table 1.1**  Symbol appointment

| Parameter | Description |
|---|---|
| $M, N$ | Node sets |
| $R^{\|M\| \times \|N\|}, \tilde{R}^{\|M\| \times \|N\|}$ | Target HiDS matrix and its low-rank approximation |
| $r_{m,n}, \tilde{r}_{m,n}$ | A single entry from $R$ and $\tilde{R}$ |
| $\Lambda, \Gamma$ | Known and unknown entry sets of $Z$. $\Lambda$ is also adopted as the training set |
| $\Omega, \Phi$ | Validation and testing sets disjoint with $\Lambda$ |
| $\Lambda(m), \Lambda(n)$ | Subsets of $\Lambda$ related to $m \in M$ and $n \in N$ |
| $f$ | Dimension of the LF space |
| $X^{\|M\| \times f}, Y^{\|N\| \times f}$ | LF matrices with $f \gg \min\{\|M\|, \|N\|\}$ |
| $x_{m,\cdot}, y_{n,\cdot}$ | Row vectors of $P$ and $Q$ |
| $X_{m,d}, y_{n,d}$ | The $d$-th element of $x_{m,\cdot}$ and $y_{n,\cdot}$ |
| $\mathbf{R}^f$ | Dimension-$f$ space of real numbers |
| $E_f$ | A $f \times f$ identity matrix |
| $\varepsilon$ | Objective function |
| $\varepsilon_{m,n}$ | Instant loss on the training instance $r_{u,i} \in \Lambda$ |
| $T$ | Maximum iteration count |
| $\tau, \sigma$ | Last update points for $x_{m,\cdot}, y_{n,\cdot}$ |
| $\eta, \lambda$ | Learning rate and regularization coefficient |
| $\kappa$ | Momentum coefficient |
| $\alpha, \beta$ | Divergence parameters |
| $u$ | Update velocity in Momentum methods |
| $k, q$ | Search dimension and swarm size |
| $s_j, v_j$ | Position and velocity of the $j$-th particle |
| $\hat{s}, \tilde{s}$ | Lower and upper bounds for the particle's position |
| $\hat{v}, \tilde{v}$ | Lower and upper bounds for the particle's velocity |
| $pb_j$ | Best position of the $j$-th particle |
| $gb$ | Best position in the whole swarm |
| $w$ | Inertia weight |
| $c_1, c_2$ | Acceleration coefficients |
| $r_1, r_2$ | Uniformly random numbers in the $(0, 1)$ interval |
| $\rho$ | Control coefficient in the range of $[0, 1]$ |
| $F(\cdot)$ | Fitness function |
| $t$ | Evolving iteration |
| $A(\cdot)$ | Quantizing function |
| $\|\cdot\|_{abs}$ | Absolute value of an enclosed number |
| $\|\cdot\|_2$ | $L_2$ norm of an enclosed vector |
| $u^{\mathrm{T}}$ | Transpose of a matrix $u$ |

## *1.2.2   An LFA Model*

Note that an HiDS matrix is defined as [1–3, 25–27]:

**Table 1.2** Abbreviation appointment

| Abbreviation | Description |
|---|---|
| HiDS | High-dimensional and sparse |
| LFA | Latent factor analysis |
| LF | Latent factor |
| PSO | Particle swarm optimization |
| $P^2SO$ | Position-transitional particle swarm optimization |
| NLFA | Non-negative latent factor analysis |
| NLF | Non-negative latent factor |
| SGD | Stochastic gradient descent |
| SLF-NMU | Single latent factor-dependent, non-negative and multiplicative update |
| SLF-NM²U | Single latent factor-dependent, non-negative, multiplicative and momentum-incorporated update |

**Fig. 1.1** An HiDS matrix

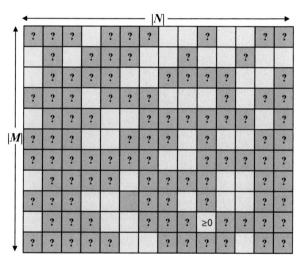

## An HDI Matrix R

*Definition 1* Given two large node sets $M$ and $N$, let $R^{|M| \times |N|}$'s each entry $r_{m,n}$ denote the interaction between nodes $m \in M$ and $n \in N$. Let $\Lambda$ and $\Gamma$ denote $R$'s known and unknown node sets, $R$ is an HiDS matrix if $|\Lambda| \gg |\Gamma|$. Such an HiDS matrix is shown in Fig. 1.1.

An LFA model tries to build a low-rank approximation to an HiDS matrix, which is defined as follows:

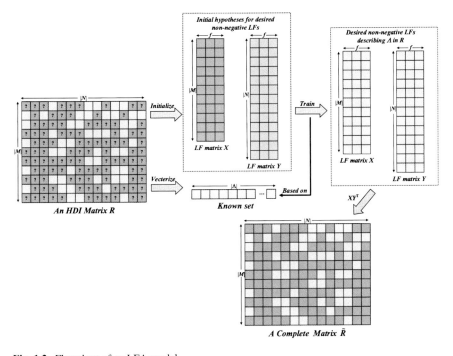

**Fig. 1.2** Flowchart of an LFA model

***Definition 2*** Given $R$ and $\Lambda$, an LFA model seeks for a rank-$f$ approximation $\tilde{R}^{|M| \times |N|}$ to $R$ based on $\Lambda$ with $\tilde{R}=XY^{\mathrm{T}}$. With the Euclidean distance, such an objective function is formulated by:

$$\varepsilon(X, Y) = \sum_{r_{m,n}} \left( r_{m,n} - \sum_{d=1}^{f} x_{m,d} y_{n,d} \right)^2 \tag{1.1}$$

where $r_{m,n}$, $x_{m,d}$ and $y_{n,d}$ denote the single elements of $R$, $X$ and $Y$, respectively.

Furthermore, in order to avoid overfitting, Tikhonov regularization [60] is introduced into (1.1). Therefore, let $\tilde{r}_{m,n} = \sum_{d=1}^{f} x_{m,d} y_{n,d}$, and then (1.1) is extended into:

$$\varepsilon(X, Y) = \sum_{r_{m,n} \in \Lambda} \left( (r_{m,n} - \tilde{r}_{m,n})^2 + \lambda \|x_{m,.}\|_2^2 + \lambda \|y_{n,.}\|_2^2 \right)$$

$$= \sum_{r_{m,n} \in \Lambda} \left( (r_{m,n} - \tilde{r}_{m,n})^2 + \lambda \sum_{d=1}^{f} x_{m,d}^2 + \lambda \sum_{d=1}^{f} y_{n,d}^2 \right) \tag{1.2}$$

where $\lambda$ denotes the regularization coefficient, $x_{m,.}$ is the $m$-th row vector in $X$, $y_{n,.}$ is the $n$-th row vector and $Y$, respectively. $\|\cdot\|_2$ calculates the $L_2$ norm of the enclosed vector. Note that in (1.2) we adopt the principle of a data-density-oriented $L_2$ norm-based regularization proposed in [24] to diversify the regularization effects on involved LFs. And then we can adopt an efficient optimization algorithm to obtain the desired LF matrices $X$ and $Y$. The flowchart of an LFA model is in Fig. 1.2.

### 1.2.3 Particle Swarm Optimization

A standard PSO is an evolutionary meta-heuristics algorithm adopted in many fields [30, 37–39]. For particle $j$, its next flying is decided by its own flying experience $pb_j$ and the best flying experience $gb$ among its peers. Hence, given a swarm consisting of $q$ particles, its evolution rule regarding the $j$-th particle at the $t$-th iteration is:

$$\forall j \in \{1, \ldots, q\} :$$
$$\begin{cases} v_j(t) = wv_j(t-1) + c_1 r_1 \big(pb_j(t-1) - s_j(t-1)\big) + c_2 r_2 \big(gb(t-1) - s_j(t-1)\big) \\ s_j(t) = s_j(t-1) + v_j(t) \end{cases}$$

$$(1.3)$$

where $s_j$ and $v_j$ are position and velocity of the $j$-th particle, $w$ is the non-negative inertia constant, $c_1$ and $c_2$ are cognitive and social coefficients, $r_1$ and $r_2$ are two uniform random numbers in the range of [0, 1], respectively.

## 1.3  Book Organization

The rest of this book is organized as follows:

- Chapter 1
    This chapter briefly reviews some background knowledge and work related to the main methodology that will be explored in this book.
- Chapter 2
    In this chapter, we propose a novel learning rate adaptation approach for an LFA model by incorporating the principle of PSO. In Phase 1, we present a latent factor learning algorithm to learn the row and column LFs of a target HiDS matrix. In Phase 2, we introduce the principle of PSO into the learning process by building a swarm of learning rates applied to the same group of LFs. Based on Phases 1 and 2, Phase 3 designs a *Learning rate-free LFA* ($L^2$FA) model. In Phase 4, we conduct extensive experiments to demonstrate the effectiveness of an $L^2$FA model.
- Chapter 3
    In this chapter, we adopt the principle of PSO to achieve a novel learning rate and regularization coefficient adaptation approach for an LFA model. We first show a latent factor learning algorithm to learn the row and column LFs of a

target HiDS matrix. After that, we build a swarm by taking the learning rate and regularization coefficient of every single LFA-based model as particles, and then apply particle swarm optimization to make them adaptation according to a pre-defined fitness function. Based on such design, a *L*earning rate and *R*egularization coefficient-free *LFA* (LRLFA) model is proposed. Experimental results indicate that an LRLFA implements efficient self-adaptation of hyper-parameters.

- Chapter 4

   In this chapter, we first build an $\alpha$-$\beta$-divergence-generalized learning objective for an LFA model with non-negativity constraint. And then, we specially designed a fast latent factor learning algorithm for this learning objective to learn the row and column LFs of a target HiDS matrix. After that, we implement self-adaptation of the regularization coefficient and momentum coefficient for excellent practicability via PSO. Based on such design, a *G*eneralized and *A*daptive *LFA* (GALFA) model is proposed. Empirical studies demonstrate that a GALFA model achieves significant accuracy and efficiency gain when compared the state-of-the-art models.

- Chapter 5

   In this chapter, we first investigate the evolution process of a standard PSO algorithm, and then propose a novel position-transitional particle swarm optimization ($P^2SO$) algorithm for avoiding its premature convergence. And then, a latent factor learning algorithm to learn the row and column LFs of a target HiDS matrix is proposed. After that, we build a swarm of learning rates applied to the same group of LFs, and adopt $P^2SO$ into the learning process to make them adaptation with high efficiency. Based on such design, an *A*dvanced *L*earning rate-free *LFA* ($AL^2FA$) model is proposed. Experimental results demonstrate that an $AL^2FA$ model's prediction accuracy and computational efficiency are highly competitive.

- Chapter 6

   The last chapter summarizes this book and provides some future directions that can be explored.

In order to make each of these chapters self-contained, some critical contents, e.g., model definitions or motivations having appeared in previous chapters, may be briefly reiterated in some chapters.

## References

1. Luo, X., Wang, D.X., Zhou, M.C., Yuan, H.Q.: Latent factor-based recommenders relying on extended stochastic gradient descent algorithms. IEEE Trans. Syst. Man Cybern. Syst. **51**(2), 916–926 (2021)
2. Luo, X., Wang, Z.D.: Assimilating second-order information for building non-negative latent factor analysis-based recommenders. IEEE Trans. Syst. Man Cybern. Syst. **52**(1), 485–497 (2021)

3. Luo, X., Zhou, M., Xia, Y., Zhu, Q.: An incremental-and-static-combined scheme for matrix-factorization-based collaborative filtering. IEEE Trans. Autom. Sci. Eng. **13**(1), 333–343 (2016)
4. Zhu, Y.C., Chen, Z.Z.: Mutually-regularized dual collaborative variational auto-encoder for recommendation systems. In: Proc. of the ACM Web Conference 2022, pp. 2379–2387. Association for Computing Machinery (2022)
5. Jin, J.C., Guo, H.F., Xu, J., Wang, X., Wang, F.Y.: An end-to-end recommendation system for urban traffic controls and management under a parallel learning framework. IEEE Trans. Intell. Transport. Syst. **22**(3), 1616–1626 (2021)
6. Xia, Y.N., Zhou, M.C., Luo, X., Zhu, Q.S., Li, J., Huang, Y.: Stochastic modeling and quality evaluation of infrastructure-as-a-service clouds. IEEE Trans. Autom. Sci. Eng. **12**(1), 162–170 (2015)
7. Sun, L., Ma, J.G., Wang, H., Zhang, Y.C., Yong, J.M.: Cloud service description model: an extension of USDL for cloud services. IEEE Trans. Serv. Comput. **11**(2), 354–368 (2018)
8. Wang, Y.C., He, Q., Zhang, X.Y., Ye, D.Y., Yang, Y.: Efficient QoS-aware service recommendation for multi-tenant service-based systems in cloud. IEEE Trans. Serv. Comput. **13**(6), 1045–1058 (2020)
9. Xia, Y.N., Zhou, M.C., Luo, X., Pang, S.C., Zhu, Q.S.: A stochastic approach to analysis of energy-aware Dvs-enabled cloud datacenters. IEEE Trans. Syst. Man Cybern. Syst. **45**(1), 73–83 (2015)
10. Wu, C., Toosi, A.N., Buyya, R., Ramamohanarao, K.: Hedonic pricing of cloud computing services. IEEE Trans. Cloud Comput. **9**(1), 182–196 (2021)
11. Wang, Q., Liu, X., Shang, T., Liu, Z., Yang, H., Luo, X.: Multi-constrained embedding for accurate community detection on undirected networks. IEEE Trans. Netw. Sci. Eng. **9**(5), 3675–3690 (2022). https://doi.org/10.1109/TNSE.2022.3176062
12. Zhang, S.S., Liang, X., Wei, Y.D., Zhang, X.: On structural features, user social behavior, and kinship discrimination in communication social networks. IEEE Trans. Comput. Soc. Syst. **7**(2), 425–436 (2020)
13. Chen, X., Proulx, B., Gong, X.W., Zhang, J.S.: Exploiting social ties for cooperative D2D communications: a mobile social networking case. IEEE/ACM Trans. Netw. **23**(5), 1471–1484 (2015)
14. Liu, S.X., Hu, X.J., Wang, S.H., Zhang, Y.D., Fang, X.W., Jiang, C.Q.: Mixing patterns in social trust networks: a social identity theory perspective. IEEE Trans. Comput. Soc. Syst. **8**(5), 1249–1261 (2021)
15. Whitaker, R.M., et al.: The coevolution of social networks and cognitive dissonance. IEEE Trans. Comput. Soc. Syst. **9**(2), 376–393 (2022)
16. Zhou, X.P., Liang, X., Du, X., Zhao, J.: Structure based user identification across social networks. IEEE Trans. Knowl. Data Eng. **30**(6), 1178–1191 (2018)
17. Yi, C.W.: A unified analytic framework based on minimum scan statistics for wireless ad hoc and sensor networks. IEEE Trans. Parallel Distrib. Syst. **20**(9), 1233–1245 (2009)
18. Zhao, C., Zhang, W.X., Yang, Y., Yao, S.: Treelet-based clustered compressive data aggregation for wireless sensor networks. IEEE Trans. Veh. Technol. **64**(9), 4257–4267 (2015)
19. Zhao, M., Li, J., Yang, Y.Y.: A framework of joint mobile energy replenishment and data gathering in wireless rechargeable sensor networks. IEEE Trans. Mobile Comput. **13**(12), 2689–2705 (2014)
20. Pang, K., Lin, Z.H., Uchoa-Filho, B.F., Vucetic, B.: Distributed network coding for wireless sensor networks based on rateless LT codes. IEEE Wireless Commun. Lett. **1**(6), 561–564 (2012)
21. Agarwal, A., Jagannatham, A.K.: Distributed estimation in homogenous Poisson wireless sensor networks. IEEE Wireless Commun. Lett. **3**(1), 90–93 (2014)
22. Quoc, D.N., Liu, N.S., Guo, D.H.: A hybrid fault-tolerant routing based on Gaussian network for wireless sensor network. J. Commun. Netw. **24**(1), 37–46 (2022)

23. Luo, X., Wu, H., Li, Z.: NeuLFT: a novel approach to nonlinear canonical polyadic decomposition on high-dimensional incomplete tensors. IEEE Trans. Knowl. Data Eng. https://doi.org/10.1109/TKDE.2021.3176466

24. Luo, X., Wang, Z., Shang, M.: An instance-frequency-weighted regularization scheme for non-negative latent factor analysis on high dimensional and sparse data. IEEE Trans. Syst. Man Cybern. Syst. **51**(6), 3522–3532 (2021)

25. Luo, X., Qin, W., Dong, A., Sedraoui, K., Zhou, M.C.: Efficient and high-quality recommendations via momentum-incorporated parallel stochastic gradient descent-based learning. IEEE/CAA J. Autom. Sin. **8**(2), 402–411 (2021)

26. Chen, J., Luo, X., Zhou, M.C.: Hierarchical particle swarm optimization-incorporated latent factor analysis for large-scale incomplete matrices. IEEE Trans. Big Data. https://doi.org/10.1109/TBDATA.2021.3090905

27. Luo, X., Zhou, M.C., Li, S., Hu, L., Shang, M.: Non-negativity constrained missing data estimation for high-dimensional and sparse matrices from industrial applications. IEEE Trans. Cybern. **50**(5), 1844–1855 (2020)

28. Song, Y., Li, M., Luo, X., Yang, G., Wang, C.: Improved symmetric and nonnegative matrix factorization models for undirected, sparse and large-scaled networks: a triple factorization-based approach. IEEE Trans. Ind. Inform. **16**(5), 3006–3017 (2020)

29. Shang, M., Luo, X., Liu, Z., Chen, J., Yuan, Y., Zhou, M.C.: Randomized latent factor model for high-dimensional and sparse matrices from industrial applications. IEEE/CAA J Autom. Sin. **6**(1), 131–141 (2019)

30. Luo, X., Zhou, M.C., Li, S., Xia, Y., You, Z., Zhu, Q., Leung, H.: Incorporation of efficient second-order solvers into latent factor models for accurate prediction of missing QoS data. IEEE Trans. Cybern. **48**(4), 1216–1228 (2018)

31. Luo, X., Sun, J., Wang, Z., Li, S., Shang, M.: Symmetric and non-negative latent factor models for undirected, high dimensional and sparse networks in industrial applications. IEEE Trans. Ind. Inform. **13**(6), 3098–3107 (2017)

32. Nishioka, Y., Taura, K.: Scalable task-parallel SGD on matrix factorization in multicore architectures. In: IEEE International Parallel and Distributed Processing Symposium Workshop, pp. 1178–1184 (May 2015)

33. Ma, H., King, I., Lyu, M.R.: Learning to recommend with social trust ensemble. In: Proc. of the 32nd Int. ACM SIGIR Conf. on Research and Development in Information Retrieval, pp. 203–210 (2009)

34. Liu, X., Yang, Y.J., Xu, Y.B., Yang, F.N., Huang, Q.Y., Wang, H.: Real-time POI recommendation via modeling long- and short-term user preferences. Neurocomputing. **467**, 454–464 (2022)

35. Hu, L., Zhang, J., Pan, X., Luo, X., Yuan, H.: An effective link-based clustering algorithm for detecting overlapping protein complexes in protein-protein interaction networks. IEEE Trans. Netw. Sci. Eng. **8**(4), 3275–3289 (2021)

36. Bi, K., Tu, K., Gu, N, et al.: Topological hole detection in sensor networks with cooperative neighbors. In: Inter. Conf. on Systems and Networks Communications, pp. 31–31 (2006)

37. Luo, X., Liu, Z., Shang, M., Lou, J., Zhou, M.C.: Highly-accurate community detection via pointwise mutual information-incorporated symmetric non-negative matrix factorization. IEEE Trans. Netw. Sci. Eng. **8**(1), 463–476 (2021)

38. Luo, X., Liu, Z., Jin, L., Zhou, Y., Zhou, M.C.: Symmetric non-negative matrix factorization-based community detection models and their convergence analysis. IEEE Trans. Neural Netw. Learn. Syst. **33**(3), 1203–1215 (2022). https://doi.org/10.1109/TNNLS.2020.3041360

39. Luo, X., Zhou, M.C., Xia, Y.N., et al.: Generating highly accurate predictions for missing QoS data via aggregating nonnegative latent factor models. IEEE Trans. Neural Netw. Learn. Syst. **27**(3), 524–537 (2015)

40. Le Nguyen, V., Caverly, R.J.: Cable-driven parallel robot pose estimation using extended Kalman filtering with inertial payload measurements. IEEE Robot. Autom. Lett. **6**(2), 3615–3622 (2021)

41. Cao, H.Q., Nguyen, H.X., Tran, T.N.-C., Tran, H.N., Jeon, J.W.: A robot calibration method using a neural network based on a butterfly and flower pollination algorithm. IEEE Trans. Ind. Electron. **69**(4), 3865–3875 (2022)
42. Chen, X., Zhan, Q.: The kinematic calibration of an industrial robot with an improved beetle swarm optimization algorithm. IEEE Robot. Autom. Lett. **7**(2), 4694–4701 (2022)
43. Yang, L., Lv, C., Wang, X., et al.: Collective entity alignment for knowledge fusion of power grid dispatching knowledge graphs. IEEE/CAA J. Autom. Sin. **9**(4), 1–15 (2022)
44. Aktosun, T., Choque-Rivero, A.E.: Factorization of the transition matrix for the general Jacobi system. Math. Methods Appl. Sci. **40**(6), 1964–1972 (2017)
45. Cantó, R., Peláez, M.J., Urbano, A.M.: Full rank Cholesky factorization for rank deficient matrices. Appl. Math. Lett. **40**, 17–22 (2015)
46. Salakhutdinov, R., Mnih, A.: Probabilistic matrix-factorization. In: Proc. of the 20th Int. Conf. on Neural Information Processing Systems, vol. 20, pp. 1257–1264 (2007)
47. Yu, K., Zhu, S.H., Lafferty, J., Gong, Y.H.: Fast nonparametric matrix factorization for large-scale collaborative-filtering. In: Proc. 32nd ACM SIGIR Conf. on Research and Development in Information Retrieval, pp. 211–218 (2009)
48. Zhang, S., Yao, L., Xu, X.: AutoSVD++: an efficient hybrid collaborative filtering model via contractive auto-encoders. In: Proc. of the 40th Int. ACM SIGIR Conf. on Research and Development in Information Retrieval, pp. 957–960 (2017)
49. Yuan, Y., He, Q., Luo, X., Shang, M.S.: A multilayered-and-randomized latent factor model for high-dimensional and sparse matrices. IEEE Trans. Big Data. **8**(3), 784–794 (2022)
50. Koren, Y., Bell, R., Volinsky, C.: Matrix factorization techniques for recommender systems. IEEE Comput. **42**(8), 30–37 (2009)
51. Chen, Y., Chen, B., He, X., Gao, C., Li, Y., Lou, J.G., Wang, Y.: λOpt: learn to regularize recommender models in finer levels. In: Proc. of the 25th ACM SIGKDD Int. Conf. on Knowledge Discovery and Data Mining, pp. 978–986 (May 2019)
52. Rendle, S.: Learning recommender systems with adaptive regularization. In: ACM, vol. 133 (2012)
53. Feng, N., Benjamin, R., Christopher, R., Stephen, J.W.: Hogwild!: a lock-free approach to parallelizing stochastic gradient descent. Adv. Neural Inf. Proces. Syst. **24**, 693–701 (2011)
54. Li, H., Li, K.L., An, J.Y., Li, K.Q.: MSGD: a novel matrix factorization approach for large-scale collaborative filtering recommender systems on GPUs. IEEE Trans. Parallel Distrib. Syst. **29**(7), 1530–1544 (2018)
55. Luo, X., Liu, Z., Li, S., Shang, M., Wang, Z.: A fast non-negative latent factor model based on generalized momentum method. IEEE Trans. Syst. Man Cybern. Syst. **51**(1), 610–620 (2021)
56. Yue, C.T., Qu, B.Y., Liang, J.: A multiobjective particle swarm optimizer using ring topology for solving multimodal multiobjective problems. IEEE Trans. Evol. Comput. **22**(5), 805–817 (2018)
57. Hu, W., Yen, G.G.: Adaptive multiobjective particle swarm optimization based on parallel cell coordinate system. IEEE Trans. Evol. Comput. **19**(1), 1–18 (2015)
58. Jiang, Y., Han, F.: A hybrid algorithm of adaptive particle swarm optimization based on adaptive moment estimation method. Intell. Comput. Theor. Appl., 658–667 (2017)
59. Yu, Z.H., Xiao, L.J., Li, H.Y., Zhu, X.L., Huai, R.T.: Model parameter identification for lithium batteries using the coevolutionary particle swarm optimization method. IEEE Trans. Ind. Electron. **64**(7), 5690–5700 (2017)
60. Luo, X., Zhou, M.-C., Li, S., Shang, M.-S.: An inherently non-negative latent factor model for high-dimensional and sparse matrices from industrial applications. IEEE Trans. Ind. Inform. **14**(5), 2011–2022 (2018)

# Chapter 2
# Learning Rate-Free Latent Factor Analysis via PSO

## 2.1 Overview

With the explosive growth of the Internet, data generated by numerous industrial applications grows exponentially. Therefore, it is difficult for people to obtain useful information from big data, which contain a wealth of knowledge and are high-dimensional and sparse (HiDS) [1–5], e.g., node interaction in sensor networks [6–8], user-service invoking in cloud computing [9–15], protein interaction in biological information [16–18], user interactions in social networks service systems [19–21], and user-item preferences in recommender systems [22–25].

An HiDS matrix is commonly adopted to describe such specific data [1–3, 10–15, 26, 27]. However, its massive unknown data obstruct people filtering core information from it precisely [1]. As shown in previous work [28, 29], a latent factor analysis (LFA) model has been proven to address such an HiDS matrix in practice owing to its high efficiency and scalability. An LFA model works by mapping a target HiDS matrix into the same and low-dimensional LF space, constructs a loss function on the observed data of an HiDS matrix and desired LFs, and then minimizes this loss function to them. To data, a large number of LFA models with good performance has been proposed, e.g., a biased regularized incremental matrix factorization model [28], a probabilistic matrix factorization model [29], a collaborative Gaussian process-based preference model [30], and a weighted trace-norm regularization-based model [31].

Stochastic gradient descent (SGD) is a common adopted optimization algorithm to perform LF analysis on an HiDS matrix [1–5]. However, the learning rate of SGD has great influence on the performance of an LFA model. For instance, small learning rate leads to slow convergence and large learning rate might result in failing to find the optimal solution. Hence, the right choice of learning rate is crucial. In general, the two most commonly used strategies to choose the value of learning rate are as follows:

© The Author(s), under exclusive license to Springer Nature Singapore Pte Ltd. 2022
Y. Yuan, X. Luo, *Latent Factor Analysis for High-dimensional and Sparse Matrices*,
SpringerBriefs in Computer Science, https://doi.org/10.1007/978-981-19-6703-0_2

a)  Grid-search. It tunes the value of learning rate on a probe dataset, and then using the resultant empirical value on the target data to design the required LF model. However, this strategy is data-depended and need to implement numerous experiments to adjust learning rate to get an empirical value. Hence, the tuning time is unacceptable.

b)  Extended SGD. Duchi et al. [32] propose AdaGrad that automatically adjusts the learning rate by performing larger updates for infrequent and smaller updates for frequent parameters. Zeiler [33] proposes AdaDelta that implements learning rate adaptation based on the decaying average of all past squared gradients. Li et al. [34] design AdaError that adjusts the learning rate based on the noisiness level of data. Although the aforementioned methods implements effective leaning rate adaptation, they make an LFA model's time cost per iteration increase significantly, thereby resulting in the increased total time cost.

Thus, we have the research question of this chapter:

*RQ:* Can we find another path to learning rate adaptation in an SGD-based LFA model without additional computation burden?

As unveiled by prior study [35–38], a particle swarm optimization (PSO) algorithm is able to make hyper-parameter adaptation in many optimization problems with the following advantages: (a) high compatibility and convergence rate; and (b) good performance in unraveling difficult optimization problems.

Inspired by this discovery, we propose a *Learning rate-free LFA* ($L^2FA$) model by incorporating the principle of PSO. This model can implement learning rate adaptation with high computational efficiency. The main idea is to introduce the principle of PSO into the learning process by building a swarm of learning rates applied to the same group of latent factors (LFs). The main contributions of this chapter include:

a)  An $L^2FA$ model. It implements efficient self-adaptation of learning rate;
b)  Algorithm design and analysis of an $L^2FA$ model.

Empirical studies on four HiDS matrices from industrial applications demonstrate the effectiveness of the proposed $L^2FA$ model. To the authors' best knowledge, such efforts have been never seen in any previous study.

The remainder of this chapter is organized as follows: Sect. 2.2 describes an SGD-based LFA model. Section 2.3 presents our $L^2FA$ model in detail. Section 2.4 introduces the experimental results. Section 2.5 concludes the chapter.

## 2.2  An LFA Model with SGD Algorithm

As shown in Sect. 1.2.2, the objective function [1–8, 28] of an LFA model is formulated by:

$$\varepsilon(X, Y) = \sum_{r_{m,n} \in \Lambda} \left( (r_{m,n} - \tilde{r}_{m,n})^2 + \lambda \|x_{m,.}\|_2^2 + \lambda \|y_{n,.}\|_2^2 \right)$$

$$= \sum_{r_{m,n} \in \Lambda} \left( (r_{m,n} - \tilde{r}_{m,n})^2 + \lambda \sum_{d=1}^{f} x_{m,d}^2 + \lambda \sum_{d=1}^{f} y_{n,d}^2 \right) \tag{2.1}$$

where $\tilde{r}_{m,n} = \sum_{d=1}^{f} x_{m,d} y_{n,d}$, $\lambda$ denotes the regularization coefficient, $x_{m,}$ is the $m$-th row vector in $X$, $y_{n,}$ is the $n$-th row vector and $Y$, $r_{m,n}$, $x_{m,d}$ and $y_{n,d}$ denote the single elements of $R$, $X$ and $Y$, $\|\cdot\|_2$ calculates the $L_2$ norm of the enclosed vector.

When performing LFA on an HiDS matrix, an SGD algorithm enjoys its fast convergence and ease of implementation. With it, (2.1) is minimized with desired LF matrices $X$ and $Y$ as follows:

$$\arg\min_{X,Y} \varepsilon(X, Y) \overset{SGD}{\Rightarrow} \forall r_{m,n} \in \Lambda, d \in \{1, 2, \ldots, f\} :$$

$$x_{m,d}^{\tau} \leftarrow x_{m,d}^{\tau-1} - \eta \frac{\partial \varepsilon_{m,n}^{\tau-1}}{\partial x_{m,d}^{\tau-1}}, y_{n,d}^{\sigma} \leftarrow y_{n,d}^{\sigma-1} - \eta \frac{\partial \varepsilon_{m,n}^{\sigma-1}}{\partial y_{n,d}^{\sigma-1}}; \tag{2.2}$$

Where $\varepsilon_{m,n} = (r_{m,n} - \tilde{r}_{m,n})^2 + \lambda_X \|x_{m,.}\|_2^2 + \lambda_Y \|y_{n,.}\|_2^2$ is the instant error corresponding to the training instance $r_{m,n} \in \Lambda$, and $\tau$ and $(\tau - 1)$ denote the current and last update points for $x_{m,d}$, $\sigma$ and $\sigma - 1$ denote the current and last update points for $y_{n,d}$, respectively. Based on (2.1), we achieve the following expressions:

$$\frac{\partial \varepsilon_{m,n}^{\tau-1}}{\partial x_{m,d}^{\tau-1}} = -err_{m,d}^{\tau-1} \cdot y_{n,d}^{\sigma-1} + \lambda x_{m,d}^{\tau-1},$$

$$\frac{\partial \varepsilon_{m,n}^{\sigma-1}}{\partial y_{n,d}^{\sigma-1}} = -err_{m,d}^{\sigma-1} \cdot x_{m,d}^{\tau-1} + y_{n,d}^{\sigma-1}; \tag{2.3}$$

where $err_{m,n} = r_{m,n} - \overset{\ddot{A}}{r}_{m,n}$, and substituting (2.3) into (2.2), we achieve the following update rules of LF matrices $X$ and $Y$ as follows:

$$\arg\min_{X,Y} \varepsilon(X, Y) \overset{SGD}{\Rightarrow} \forall r_{m,n} \in \Lambda, d \in \{1, 2, \ldots, f\} :$$

$$\begin{cases} x_{m,d}^{\tau} \leftarrow x_{m,d}^{\tau-1} + \eta \left( err_{m,d}^{\tau-1} \cdot y_{n,d}^{\sigma-1} - \lambda x_{m,d}^{\tau-1} \right), \\ y_{n,d}^{\sigma} \leftarrow y_{n,d}^{\sigma-1} + \eta \left( err_{m,d}^{\sigma-1} \cdot x_{m,d}^{\tau-1} - y_{n,d}^{\sigma-1} \right). \end{cases} \tag{2.4}$$

From (2.4), we can find that the learning rate $\eta$ determine the performance of an LFA model with SGD. Hence, how to find the appropriate learning rate is our concern.

## 2.3   The Proposed L²FA Model

In this section, a novel L²FA model is proposed. It implements efficient self-adaptation of learning rate by incorporating the principle of PSO into the learning process.

### 2.3.1   Learning Rate Adaptation via PSO

To implement efficient self-adaptation of learning rate in an SGD-based LFA model, we define that each particle in the swarm denotes a separate learning rate $\eta$. As shown in Sect. 1.2.3, the evolution rule regarding the $j$-th learning rate at the $t$-th iteration is:

$$\forall j \in \{1, \ldots, q\}:$$

\scale90%
$$\begin{cases} v_j(t) = wv_j(t-1) + c_1 r_1 \left(pb_j(t-1) - s_j(t-1)\right) + c_2 r_2 \left(gb(t-1) - s_j(t-1)\right) \\ \eta_j(t) = \eta_j(t-1) + v_j(t) \end{cases}$$

$$(2.5)$$

Where $\eta_j$ and $v_j$ are position and velocity of the $j$-th learning rate, $pb_j$ and $gb$ denotes the best particle location of particle and the best position of the whole swarm, $w$ is the non-negative inertia constant, $c_1$ and $c_2$ are cognitive and social coefficients, $r_1$ and $r_2$ are two uniform random numbers in the range of [0, 1], respectively. Note that in this chapter, only one hyper-parameter needs to be adaptation. Hence, the particles are searched in a one-dimensional space only.

Note that all the learning rates linked to the particles should be restricted in a pre-defined range:

$$v_j(t) = \begin{cases} \check{v}, & v_j(t) > \check{v} \\ \hat{v}, & v_j(t) < \hat{v} \end{cases},$$

$$\eta_j(t) = \begin{cases} \check{\eta}, & \eta_j(t) > \check{\eta} \\ \hat{\eta}, & \eta_j(t) < \hat{\eta} \end{cases}. \qquad (2.6)$$

where $\hat{s}$ and $\check{s}$ determines the boundary of $\eta$, $\hat{v}$ and $\check{v}$ determines the boundary of $v$, respectively. In our study, we adopt the empirical values of $\check{v} = 1$, $\hat{v} = -1$, $\check{s} = 2^{-8}$ and $\check{s} = 2^{-12}$, respectively.

Naturally, it is also essential to choose an appropriate fitness function as a criterion for the performance of PSO. Note that this chapter adopts the missing data estimation of an HiDS matrix as the representation learning objective. Hence, root mean squared error (RMSE) and mean absolute error (MAE) are adopted as the fitness functions:

$$F_1 = \sqrt{\left( \sum_{r_{m,n}\in\Omega} (r_{m,n} - \tilde{r}_{m,n})^2 \right) / |\Omega|},$$

$$F_2 = \left( \sum_{r_{m,n}\in\Omega} |r_{m,n} - \tilde{r}_{m,n}|_{abs} \right) / |\Omega|; \tag{2.7}$$

where $\Omega$ means the validation set and is disjoint with the training set $\Lambda$, $\overset{\ddot{A}}{r}_{m,n}$ is the rating estimate to $r_{m,n}$ generated by an ALF model, $|\cdot|$ computes the cardinality of a given set, and $|\cdot|_{abs}$ computes the absolute value of a given number, respectively.

Note that $\forall j \in \{1,\dots,q\}$, $\eta_j$ is linked with the same group of LF matrices, i.e., $X$ and $Y$, which are trained by SGD. If different LF matrices are selected for different particles, then connections between particles are not strong enough to form an efficient swarm. Thus, based on (2.4), its $t$-th iteration actually consists of $q$ sub-iterations, where $X$ and $Y$ are updated in the $j$-th sub-iteration as follows:

$$\arg\min_{X,Y} \varepsilon(X, Y) \overset{SGD}{\Rightarrow} \forall r_{m,n} \in \Lambda, d \in \{1, 2, \dots, f\} :$$

$$\begin{cases} x_{m,d}^\tau \leftarrow x_{m,d}^{\tau-1} + \eta\left(err_{m,d}^{\tau-1} \cdot y_{n,d}^{\sigma-1} - \lambda x_{m,d}^{\tau-1}\right), \\ y_{n,d}^\sigma \leftarrow y_{n,d}^{\sigma-1} + \eta\left(err_{m,d}^{\sigma-1} \cdot x_{m,d}^{\tau-1} - y_{n,d}^{\sigma-1}\right). \end{cases} \tag{2.8}$$

where the footnote $(j)$ on $x_{m,d}$, $y_{n,d}$ and $err_{m,n}$ denotes that their current updates are implemented with the $j$-th particle, i.e., $\eta_j$. Based on (2.5)–(2.8), we achieve an L²FA model with self-adaptive learning rate.

## 2.3.2   Algorithm Design and Analysis

On the basis of the above, we design the Algorithm L²FA. In each iteration, the algorithm traverses on $q$ and $\Lambda$, and updates involved decision parameters by using (2.5)–(2.8). Note that $V$ and $\eta$ represent the velocity and position of particles in Algorithm L²FA, $temp_j$ denotes the value of fitness function $F$ corresponding to $j$-th particle. Hence, we can obtain the computational cost of Algorithm L²FA:

$$T_{ALF} = \Theta(((|M| + |N|) \times f + q + 2q \times D + C \times (q \times |\Lambda| \times 3f + 14q))$$

$$\approx \Theta(|\Lambda| \times C \times q \times f) \tag{2.9}$$

Note that (2.9) adopts the condition $|\Lambda| \ll \max\{|M|, |N|\}$, which is implemented continuously in various industrial applications to decrease the lower-order-terms. From (2.9), we can see that the computational complexity of L$^2$FA is linear to the known set $|\Lambda|$ of an HiDS matrix.

In terms of Algorithm L$^2$FA's storage complexity, it relies on two factors: (a) an SGD-based LF model's caches $X^{|M| \times d}$ and $Y^{|N| \times d}$, whose storage cost comes to $\Theta((|M| + |N|) \times f)$; (b) PSO's auxiliary arrays $pb^q$, $V^q$ and $S^q$, whose storage cost comes to $\Theta(3q)$. Thus, by combining the above mentioned factors, and reasonably ignoring constant coefficient, we formulate the storage complexity of L$^2$FA:

$$S_{ALF} = (|M| + |N|) \times f + S + 2q \approx (|M| + |N|) \times f + q \qquad (2.10)$$

From (2.10), we can see that the storage cost of L$^2$FA is linear with the number of involved entities and swarm size. Such storage cost is easy to resolve in industrial applications.

Based on the above-mentioned inference, we can find that L$^2$FA is greatly efficient in terms of computation and storage. In next section, we verify its performance on four HiDS matrices generated by real applications.

| Algorithm L$^2$FA | |
|---|---|
| **Input:** $\Lambda$, $\Gamma$, $M$, $N$, $f$ | |
| **Operation** | Cost |
| **Initialize** $X^{|M| \times f}$, $Y^{|N| \times f}$ | $\Theta(|M| + |N|) \times f$ |
| **Initialize** $\lambda$, $q$, $r_1$, $r_2$, $c_1$, $c_2$, $w$, $gb$, $v_{max}$, $v_{min}$, $\eta_{max}$, $\eta_{min}$, $t = 1$, $T =$ Maximum Round Count | $\Theta(1)$ |
| **Initialize** $pb^q$, $V^q$, $\eta^q$ | $\Theta(q)$ |
| **while not** converge **and** $t \leq T$ **do** | $\times C$ |
|   **for** each $j \in q$ | $\times q$ |
|     **for** each $r_{m,n} \in \Lambda$ | $\times |\Lambda|$ |
|       $\tilde{r}_{m,n} = \sum_{d=1}^{f} x_{m,d} y_{n,d}$ | $\Theta(f)$ |
|       $err = r_{m,n} - \tilde{r}_{m,n}$ | $\Theta(1)$ |
|       **for** $d = 1$ **to** $f$ | $\times f$ |
|         $x_{m,d} = x_{m,d} + \eta_j \times (err \times y_{n,d} - \lambda \times x_{m,d})$ | $\Theta(1)$ |
|         $y_{n,d} = y_{n,d} + \eta_j \times (err \times x_{m,d} - \lambda \times y_{n,d})$ | $\Theta(1)$ |
|       **end for** | – |
|     **end for** | – |
|     $temp_j = F$ according to (2.7) | $\Theta(1)$ |
|   **end for** | – |
|   **for** each $j \in q$ | $\times q$ |
|     **if** $temp_j < pb_j$ | $\Theta(1)$ |
|       $pb_j = temp_j$ | $\Theta(1)$ |
|     **end if** | – |
|     **if** $temp_j < gb$ | $\Theta(1)$ |
|       $gb = temp_j$ | $\Theta(1)$ |

(continued)

| Algorithm L$^2$FA | |
|---|---|
| **end if** | – |
| **end for** | – |
| **for** each $j \in q$ | $\times q$ |
| $v_j = wv_j + c_1 r_1 (pb_j - \eta_j) + c_2 r_2 (gb - \eta_j)$ | $\Theta(1)$ |
| **if** $v_j > v_{max}$ | $\Theta(1)$ |
| $v_j = v_{max}$ | $\Theta(1)$ |
| **else if** $v_j < v_{min}$ | $\Theta(1)$ |
| $v_j = v_{min}$ | $\Theta(1)$ |
| **end if** | – |
| $\eta_j = \eta_j + v_j$ | $\Theta(1)$ |
| **if** $\eta_j > \eta_{max}$ | $\Theta(1)$ |
| $\eta_j = \eta_{max}$ | $\Theta(1)$ |
| **else if** $\eta_j < \eta_{min}$ | $\Theta(1)$ |
| $\eta_j = \eta_{min}$ | $\Theta(1)$ |
| **end if** | – |
| **end for** | – |
| $t = t + 1$ | $\Theta(1)$ |
| **end while** | – |
| **Output:** $X, Y$ | |

## 2.4   Experimental Results and Analysis

### 2.4.1   General Settings

**Evaluation Metrics**  For practical applications, we mainly use the observed data in an HiDS matrix to predict the unknown terms, so as to restore the complete relationship between entities [8, 39–41]. Therefore, we adopt the root mean squared error (RMSE) and mean absolute error (MAE) to measure the prediction accuracy of an HiDS matrix missing data:

$$RMSE = \sqrt{\left( \sum_{r_{m,n} \in \Phi} (r_{m,n} - \tilde{r}_{m,n})^2 \right) / |\Phi|},$$

$$MAE = \left( \sum_{r_{m,n} \in \Phi} |r_{m,n} - \tilde{r}_{m,n}|_{abs} \right) / |\Phi|;$$

where $\overset{\ddot{A}}{\tilde{r}}_{m,n}$ is the predictive rating generated by the testing instance $r_{m,n} \in \Phi$, $|\cdot|_{abs}$ denotes the absolute value of an enclosed number, $|\cdot|$ calculates the cardinality of an enclosed set.

**Table 2.1** Datasets details

| No. | Name | Row | Column | Known Entries | Density (%) |
|-----|------|-----|--------|---------------|-------------|
| D1 | Douban | 58,541 | 129,490 | 16,830,839 | 0.22 |
| D2 | EachMovie | 72,916 | 1628 | 2,811,718 | 2.37 |
| D3 | Epinion | 120,492 | 775,760 | 13,668,320 | 0.015 |
| D4 | MovieLens 10M | 10,681 | 71,567 | 10,000,054 | 1.31 |

**Table 2.2** The optimal learning rate for M1

| Dataset | $\eta$ |
|---------|--------|
| D1 | $2^{-12}$ |
| D2 | $2^{-8}$ |
| D3 | $2^{-10}$ |
| D4 | $2^{-12}$ |

**Datasets** Four HiDS matrices are included in our experiments, which is included in Table 2.1. In order to verify the feasibility of the ALF model in addressing big data from real applications, all datasets are real datasets collected by industrial companies.

Note that the known entry set of each HiDS matrix is randomly split into ten disjoint and equally-sized subsets, where eight subsets are chosen as the training set, and the remaining as the testing set for train-test settings. The above process is sequentially repeated five times for fivefold cross-validation. The termination condition is uniform for all involved models, i.e., the iteration threshold is 1000, error threshold is $10^{-5}$, and a model's training process terminates if either threshold is met.

**Model Settings** Two models are involved in the experiments. We call the SGD-based LF model as M1, and the proposed $L^2FA$ model as M2, respectively. For obtaining objective results, following general settings are applied:

(a) For both models, we set regularization coefficient $\lambda = 0.03$.
(b) We set the dimension of the LF space $f = 20$ uniformly.
(c) For PSO in M2, we set the swarm size $q = 10$, the range of learning rate is $[2^{-12}, 2^{-8}]$, and the range of velocity is $[-1, 1]$, $c_1 = c_2 = 2$.

### 2.4.2   Performance Comparison

In this part of experiments, we compare the proposed $L^2FA$ model with an SGD-based LFA model in terms of prediction accuracy and computational efficiency.

For M1, we choose the optimal learning rate for each dataset via grid-search, which is recorded in Table 2.2. The comparison results corresponding to RSME,

**Table 2.3** Lowest RMSE of M1 and M2

| Dataset | Lowest RMSE/training epochs | |
|---|---|---|
| | M1 | M2 |
| D1 | **0.7152**/1000 | 0.7156/**13** |
| D2 | 0.2317/673 | **0.2285/199** |
| D3 | **0.5940**/814 | 0.5948/**27** |
| D4 | **0.7831**/1000 | 0.7866/**10** |

**Table 2.4** Total time-comsuption corresponding to Table 2.3 (s)

| Dataset | M1 | M2 |
|---|---|---|
| D1 | 1462 | **189** |
| D2 | **299** | 633 |
| D3 | 983 | **300** |
| D4 | 813 | **80** |

**Table 2.5** Lowest MAE of M1 and M2

| Dataset | Lowest MAE/training epochs | |
|---|---|---|
| | M1 | M2 |
| D1 | **0.5590**/1000 | 0.5592/**13** |
| D2 | 0.1825/589 | **0.1796/180** |
| D3 | **0.3025**/936 | 0.3056/**35** |
| D4 | **0.6027**/1000 | 0.6054/**10** |

**Table 2.6** Total time-consumption corresponding to Table 2.5 (s)

| Dataset | M1 | M2 |
|---|---|---|
| D1 | 1645 | **174** |
| D2 | **304** | 561 |
| D3 | 1189 | **427** |
| D4 | 904 | **82** |

MAE, iterations and total time cost are summarized in Tables 2.3, 2.4, 2.5, and 2.6. From the results of these experiments, we obtain the following findings.

(a) An $L^2FA$ model implements efficient self-adaptation of learning rate with slight prediction accuracy loss. For instance, as shown in Table 2.3, on D1, M2 achieves the lowest RSME 0.7156, only 0.06% higher than that of M1 at 0.7152. The similar conclusions are found on D3 and D4. However, the situation is different on D2, M2 achieves the highest prediction accuracy for missing data on D2.

(b) An $L^2FA$ model's convergence rate is very fast. For instance, as summarized in Table 2.3, on D1, M2 only takes 13 iterations to converge. On the contrary, M1 takes 1000 iterations to achieve the lowest RMSE. This means M2 reduces 98.7% iterations than of M1 does. Similar situations are also observed on D2–D4.

(c) An $L^2FA$ model's computational efficiency is high. For instance, on D1, we see that M1 takes 1462 s to achieve its lowest RMSE, yet M2 only takes 189 s from Table 2.4. M2's computational efficiency is about 7.73 times of M1. Similar situations can be also encountered on D3 and D4. On D2, the situation is different. As shown in Table 2.4, we find that M2's total time cost is higher than M1. However, when considering the time cost of a learning model, its parameter tuning should be taken into consideration. Tables 2.4 and 2.6 only record the training time of M1 with the pre-tuned learning rate, but does not record the time cost to tune it. Note that the grid-search tuning is much more expensive in time than model training.

### 2.4.3   Effect of Swarm Size

The swarm size is essential for PSO, which is able to affect accuracy and computational efficiency. Therefore, the performance of $L^2FA$ is also affected by swarm size. In this experiment, we aim to validate swarm size's effects on HFLF and we set swarm size $q = \{10, 20, 40, 80, 120, 160\}$. Figures 2.1 and 2.2 depict the effect of swarm size on prediction accuracy. With different value of $q$, the results on iterations are displayed in Figs. 2.3 and 2.4, and total time-consumption is displayed in Figs. 2.5 and 2.6.

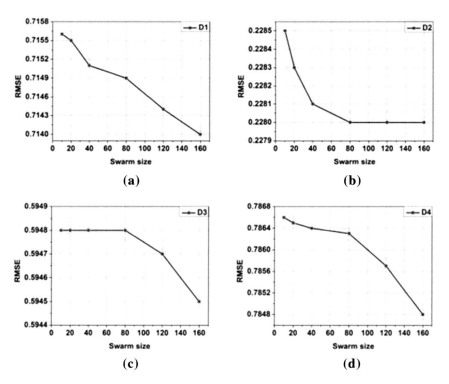

**Fig. 2.1** M2's RMSE as swarm size varies

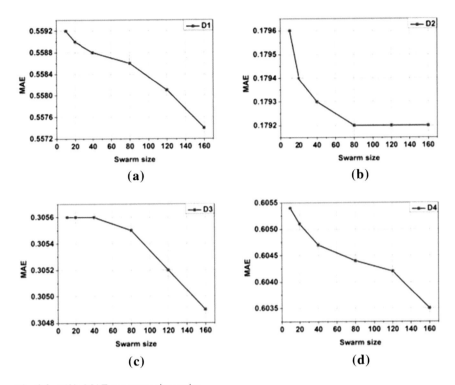

**Fig. 2.2** M2's MAE as swarm size varies

(a) An $L^2FA$ model's prediction accuracy increases as swarm size increases. For instance, in Fig. 2.1a, M2's RMSE on D1 is 0.7156, 0.7155, 0.7151, 0.7149, 0.7144 and 0.7140 when $q$ increases from 10 to 160, respectively. When $q = 160$, M2 has the lowest RMSE at 0.7140. The same situation is found on MAE. This phenomenon is reasonable since the selection of swarm size is related to the actual problem. Larger swarm size means greater chance to find a better solution.

(b) An $L^2FA$ model's iterations decreases as swarm size increases. For instance, on D1, from Fig. 2.3a, M2 takes 13 iterations with $q = 10$, 4 iterations with $q = 40$, and 2 iterations with $q = 160$. The same results can be observed on D2–D4. The main reason is that the larger swarm size means that more particles' interactive in one iterations, thereby improving the total iterations.

(c) An $L^2FA$ model's total time cost increases as swarm size increases. For instance, as shown in Fig. 2.5a, on D1, the total time cost of M2 is 189, 206, 234, 325, 346 and 517 s when $q$ increases from 10 to 160, respectively. Similar situations are also encountered on D2–D4. Since each particles in a swarm needs to carry out an SGD operation.

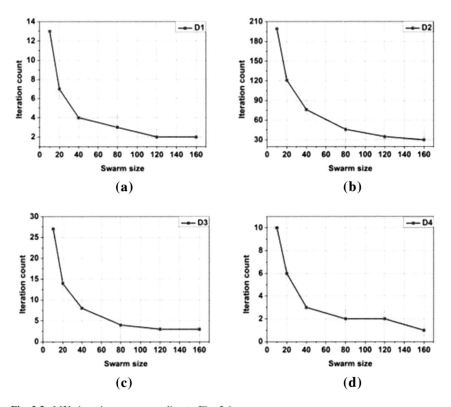

**Fig. 2.3** M2's iterations corresponding to Fig. 2.1

### 2.4.4   Summary

According to the results of experiment, we summarize that owing to its incorporation of the PSO principle into its training process, an $L^2FA$ model achieves competive prediction accuracy and high computational efficiency compared with an SGD-based LF model. Moreover, an $L^2FA$ model makes learning rate adaptation within only one full training process, which greatly improves the practicability in industrial applications.

## 2.5   Conclusions

This work adopts a standard PSO to implement learning rate adaptation for an LFA model. Although it has high computational efficiency, it may suffer from slight accuracy loss caused by premature convergence. To date, many scholars have made great efforts to overcome such a limitation [35–37, 42–44]. Moreover, the

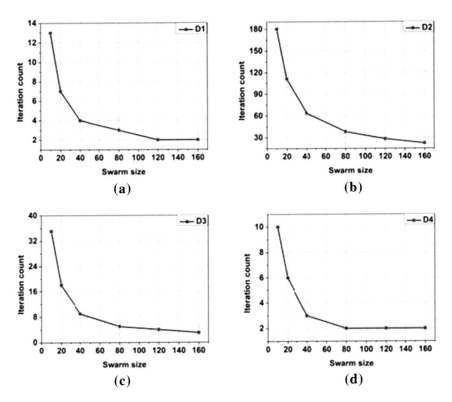

**Fig. 2.4** M2's iterations corresponding to Fig. 2.2

regularization coefficient is also affects an LFA model's prediction accuracy [45–48]. Hence, the following further study regarding these issues is planned in our future work: (a) aiming to integrate advanced PSO for further improving an LFA model's performance; and (b) Highly desired to make regularization coefficient adaptation.

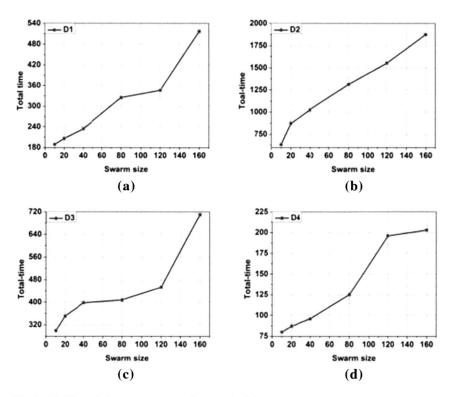

**Fig. 2.5** M2's total time cost corresponding to Fig. 2.1

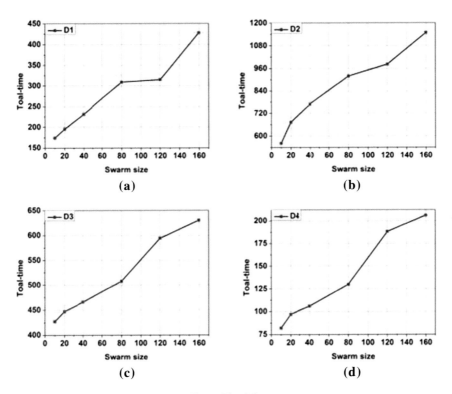

**Fig. 2.6** M2's total time cost corresponding to Fig. 2.2

# References

1. Chen, J., Luo, X., Zhou, M.C.: Hierarchical particle swarm optimization-incorporated latent factor analysis for large-scale incomplete matrices. IEEE Trans. Big Data. https://doi.org/10.1109/TBDATA.2021.3090905
2. Rafael, D., Bonifacio, M., Nicolas, M., Julian, F.: Computational intelligence tools for next generation quality of service management. Neurocomputing. **72**(16–18), 3631–3639 (2009)
3. Luo, X., Zhou, M.C., Li, S., You, Z.H., Xia, Y.N., Zhu, Q.-S.: A non-negative latent factor model for large-scale sparse matrices in recommender systems via alternating direction method. IEEE Trans. Neural Netw. Learn. Syst. **27**(3), 524–537 (2016)
4. Qian, X., Feng, H., Zhao, G., Mei, T.: Personalized recommendation combining user interest and social circle. IEEE Trans. Knowl. Data Eng. **26**(7), 1763–1777 (2014)
5. Luan, W., Liu, G., Jiang, C., Qi, L.: Partition-based collaborative tensor factorization for poi recommendation. IEEE/CAA J. Autom. Sin. **4**(3), 437–446 (2017)
6. Piao, X., Hu, Y., Sun, Y., Yin, B., Gao, J.: Correlated spatio-temporal data collection in wireless sensor networks based on low rank matrix approximation and optimized node sampling. Sensors. **14**(12), 23137–23158 (2014)
7. Nguyen, T.L., Shin, Y.: Matrix completion optimization for localization in wireless sensor networks for intelligent IoT. Sensors. **16**(5), 722 (2016)
8. Jin, L., Zhang, J.Z., Luo, X., Liu, M., Li, S., Xiao, L., Yang, Z.H.: Perturbed manipulability optimization in a distributed network of redundant robots. IEEE Trans. Ind. Electron. **68**(8), 7209–7220 (2021)

9. You, Z.H., Zhou, M.C., Luo, X., Li, S.: Highly efficient framework for predicting interactions between proteins. IEEE Trans. Cybern. **64**(6), 4710–4720 (2017)
10. Luo, X., Wu, H., Zhou, M.C., Yuan, H.Q.: Temporal pattern-aware QoS prediction via biased non-negative latent factorization of tensors. IEEE Trans. Cybern. **50**(5), 1798–1809 (2020)
11. Luo, X., Chen, M.Z., Wu, H., Liu, Z.G., Yuan, H.Q., Zhou, M.C.: Adjusting learning depth in non-negative latent factorization of tensors for accurately modeling temporal patterns in dynamic QoS data. IEEE Trans. Autom. Sci. Eng. **18**(4), 2142–2155 (2021). https://doi.org/10.1109/TASE.2020.3040400
12. Wu, D., Luo, X., Shang, M.S., He, Y., Wang, G.Y., Wu, X.D.: A data-characteristic-aware latent factor model for web services QoS prediction. IEEE Trans. Knowl. Data Eng. **34**(6), 2525–2538 (2022). https://doi.org/10.1109/TKDE.2020.3014302
13. Luo, X., Zhou, M.C., Wang, Z.D., Xia, Y.N., Zhu, Q.S.: An effective QoS estimating scheme via alternating direction method-based matrix factorization. IEEE Trans. Serv. Comput. **12**(4), 503–518 (2019)
14. Wu, D., He, Q., Luo, X., Shang, M.S., He, Y., Wang, G.Y.: A posterior-neighborhood-regularized latent factor model for highly accurate web service QoS prediction. IEEE Trans. Serv. Comput. **15**(2), 793–805 (2022). https://doi.org/10.1109/TSC.2019.2961895
15. Luo, X., Zhou, M.C., Xia, Y.N., Zhu, Q.S., Ammari, A.C., Alabdulwahab, A.: Generating highly accurate predictions for missing QoS-data via aggregating non-negative latent factor models. IEEE Trans. Neural Netw. Learn. Syst. **27**(3), 524–537 (2016)
16. Hu, L., Zhang, J., Pan, X.Y., Luo, X., Yuan, H.Q.: An effective link-based clustering algorithm for detecting overlapping protein complexes in protein-protein interaction networks. IEEE Trans. Network Sci. Eng. **8**(4), 3275–3289 (2021)
17. Hu, L., Yang, S.C., Luo, X., Yuan, H.Q., Zhou, M.C.: A distributed framework for large-scale protein-protein interaction data analysis and prediction using MapReduce. IEEE/CAA J. Autom. Sin. **9**(1), 160–172 (2022). https://doi.org/10.1109/JAS.2021.1004198
18. Hu, L., Yuan, X.H., Liu, X., Xiong, S.W., Luo, X.: Efficiently detecting protein complexes from protein interaction networks via alternating direction method of multipliers. IEEE/ACM Trans. Comput. Biol. Bioinform. **16**(6), 1922–1935 (2019)
19. Narayanam, R., Narahari, Y.: A shapley value-based approach to discover influential nodes in social networks. IEEE Trans. Autom. Sci. Eng. **8**(1), 130–147 (2010)
20. Cao, X., Wang, X., Jin, D., Cao, Y., He, D.: Erratum: identifying overlapping communities as well as hubs and outliers via nonnegative matrix factorization. Sci. Rep. **3**(10), 2993 (2014)
21. Wang, Z., Liu, Y., Luo, X., Wang, J.J., Gao, C., Peng, D.Z., Chen, W.: Large-scale affine matrix rank minimization with a novel nonconvex regularizer. IEEE Tran. Neural Netw. Learn. Syst. **33**(9), 4661–4675 (2022). https://doi.org/10.1109/TNNLS.2021.3059711
22. Gruson, F., Moigne, P.L., Delarue, P., Videt, A., Cimetiére, X., Arpillière, M.: A simple carrier-based modulation for the SVM of the matrix converter. IEEE Trans. Ind. Informat. **9**(2), 947–956 (2013)
23. Wu, D., Luo, X., Shang, M., He, Y., Wang, G.Y., Zhou, M.C.: A deep latent factor model for high-dimensional and sparse matrices in recommender systems. IEEE Trans. Syst. Man Cybern. Syst. **51**(7), 4285–4296 (2021)
24. Luo, X., Zhou, Y., Liu, Z.G., Zhou, M.C.: Fast and accurate non-negative latent factor analysis on high-dimensional and sparse matrices in recommender systems. IEEE Trans. Knowl. Data Eng. https://doi.org/10.1109/TKDE.2021.3125252
25. Wu, D., Shang, M.S., Luo, X., Wang, Z.D.: An $L_1$-and-$L_2$-norm-oriented latent factor model for recommender systems. IEEE Trans. Neural Netw. Learn. Syst. https://doi.org/10.1109/TNNLS.2021.3071392
26. Shao, H., Zheng, G.: Convergence analysis of a back-propagation algorithm with adaptive momentum. Neurocomputing. **74**(5), 749–752 (2011)
27. Konstan, J.A., Miller, B.N., Maltz, D., Herlocker, J.L., Gordon, L.R., Riedl, J.: GroupLens: applying collaborative filtering to usenet news. Commun. ACM. **40**(3), 77–87 (1997)

28. Takács, G., Pilászy, I., Németh, B., Tikky, D.: Scalable collaborative filtering approaches for large recommender systems. J. Mach. Learn. Res. **10**, 623–656 (2009)
29. Salakhutdinov, R., Mnih, A.: Probabilistic matrix factorization. Adv. Neural Inf. Proces. Syst. **20**, 1257–1264 (2008)
30. Houlsby, N., Huszar, F., Ghahramani, Z., Hernández-lobato, J.M.: Collaborative Gaussian processes for preference learning. Adv. Neural Inf. Proces. Syst., 2105–2113 (2012)
31. Srebro, N., Salakhutdinov, R.: Collaborative filtering in a non-uniform world: learning with the weighted trace norm. Adv. Neural Inf. Process. Syst., 2056–2064 (2010)
32. Duchi, J., Hazan, E., Singer, Y.: Adaptive subgradient methods for online learning and stochastic optimization. J. Mach. Learn. Res. **12**(7), 2121–2159 (2011)
33. Zeiler, M.D.: ADADELTA: an adaptive learning rate method. Comput. Sci. (2012)
34. D.-S. Li, C. Chen, Q. Lv, H.-S. Gu, T. Lu, L. Shang, N. Gu, S.-M. Chu: AdaError: an adaptive learning rate method for matrix approximation-based collaborative filtering. In: Proc. Of the 27th World Wide Web Conference, pp. 741–751 (2018)
35. Zeng, N.-Y., Hung, Y.-S., Li, Y.-R., Du, M.: A novel switching local evolutionary PSO for quantitative analysis of lateral flow immunoassay. Expert Syst. Appl. **41**(4), 1708–1715 (2014)
36. Zeng, N.-Y., Wang, Z.-D., Zhang, H., Alsaadi, F.-E.: A novel switching delayed PSO algorithm for estimating unknown parameters of lateral flow immunoassay. Cogn. Comput. **8**(2), 143–152 (2016)
37. Zeng, N.-Y., Wang, Z.-D., Li, Y.-R., Du, M., Liu, X.-H.: Identification of nonlinear lateral flow immunoassay state-space models via particle filter approach. IEEE Trans. Nanotechnol. **1**(2), 321–327 (2012)
38. Hu, L., Pan, X., Tang, Z., Luo, X.: A fast Fuzzy clustering algorithm for complex networks via a generalized momentum method. IEEE Trans. Fuzzy Syst. **30**(9), 3473–3485 (2022). https://doi.org/10.1109/TFUZZ.2021.3117442
39. Li, W., Luo, X., Yuan, H., Zhou, M.C.: A momentum-accelerated Hessian-vector-based latent factor analysis model. IEEE Trans. Serv. Comput. https://doi.org/10.1109/TSC.2022.3177316
40. Li, Z., Li, S., Luo, X.: An overview of calibration technology of industrial robots. IEEE/CAA J. Autom. Sin. **8**(1), 23–36 (2021)
41. Peng, Q., Xia, Y., Zhou, M.C., Luo, X., Wang, S., Wang, Y., Wu, C., Pang, S., Lin, M.: Reliability-aware and deadline-constrained mobile service composition over opportunistic networks. IEEE Trans. Autom. Sci. Eng. **18**(3), 1012–1025 (2021)
42. Shi, Y.-H., Eberhart, R.-C.: Parameter selection in particle swarm optimization. In: Proc. of the 7th Int. Conf. on Evolutionary Programming, pp. 591–600 (1998)
43. Ratnaweera, A., Halgamure, S.-K., Watson, H.-C.: Self-organizing hierarchical particle swarm optimizer with time-varying acceleration coefficients. IEEE Trans. Evol. Comput. **8**(3), 240–255 (2004)
44. Clerc, M., Kennedy, J.: The particle swarm: explosion, stability, and convergence in a multi-dimensional complex space. IEEE Trans. Evol. Comput. **6**(1), 58–73 (2002)
45. Wang, Q., Liu, X., Shang, T., Liu, Z., Yang, H., Luo, X.: Multi-constrained embedding for accurate community detection on undirected networks. IEEE Trans. Netw. Sci. Eng. **9**(5), 3675–3690 (2022). https://doi.org/10.1109/TNSE.2022.3176062
46. Jin, L., Zheng, X., Luo, X.: Neural dynamics for distributed collaborative control of manipulators with time delays. IEEE/CAA J. Autom. Sin. **9**(5), 854–863 (2022). https://doi.org/10.1109/JAS.2022.1005446
47. Li, Z., Li, S., Bamasag, O., Alhothali, A., Luo, X.: Diversified regularization enhanced training for effective manipulator calibration. IEEE Trans. Neural Netw. Learn. Syst. https://doi.org/10.1109/TNNLS.2022.3153039
48. Jin, L., Liang, S., Luo, X., Zhou, M.: Distributed and time-delayed K-winner-take-all network for competitive coordination of multiple robots. IEEE Trans. Cybern. https://doi.org/10.1109/TCYB.2022.3159367

# Chapter 3
# Learning Rate and Regularization Coefficient-Free Latent Factor Analysis via PSO

## 3.1 Overview

A big-data-related application commonly has numerous nodes, e.g., users and items in a recommender system [1–5]. With the exponential growth of involved nodes, it is impossible to obtain their whole interaction relationships, e.g., it is impossible for a recommender system's individual user to touch all of its items [6, 7]. For instance, the Epinions Matrix [8] is a typical HiDS matrix. It contains 631,064 ratings by 51,670 users on 83,509 different entries, with a data density of 0.015% only. Hence, the resultant interactions can be described by a high-dimensional and sparse (HiDS) matrix [9–18].

Despite its sparsity, an HiDS matrix contains rich information like the user community [19] and users' potential favorites [20]. According to the previous work [21–27], latent factor analysis (LFA) models have been proven to address such an HiDS matrix efficiently. An LFA model maps the rows and columns of an HiDS matrix into the same low-dimensional latent factor (LF) space, and builds the objective functions only based on the known values of the target matrix and desired LFs. Then it minimizes the objective function to train LFs and then predict the missing values in an HiDS matrix according to these obtained LFs.

To build an LF model, an efficiently and commonly used strategy is stochastic gradient descent (SGD), which is simplicity and low computational complexity [28–30]. However, the success of such an LFA model depends largely on the right choice of learning rate and regularization coefficient. For instance, small learning rate leads to slow convergence and large learning rate might result in failing to find the optimal solution [31, 32]. For regularization coefficient, inappropriate values make model overfitting or not learning, which both lead to poor prediction accuracy [33]. Hence, the right choice of learning rate and regularization coefficient is crucial. Based on the above analysis, we show a motivating example in Fig. 3.1.

*Motivating Example* The hyper-parameters (learning rate and regularization coefficient) have significant effects on the performance of an LFA model. For instance,

© The Author(s), under exclusive license to Springer Nature Singapore Pte Ltd. 2022
Y. Yuan, X. Luo, *Latent Factor Analysis for High-dimensional and Sparse Matrices*,
SpringerBriefs in Computer Science, https://doi.org/10.1007/978-981-19-6703-0_3

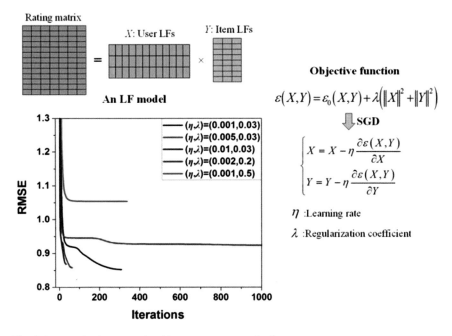

**Fig. 3.1**  A motivating example of hyper-parameters selection

*as shown in Fig. 3.1, on dataset ML1M, an LFA model consumes 310 iterations to achieve the lowest RMSE 0.8528 with η = 0.001 and λ = 0.03. However, with η = 0.002 and λ = 0.2, an LFA model consumes more than 1000 iterations to achieve the lowest RMSE 0.9239. To our surprise, different choices of hyper-parameters lead to significant fluctuation on the performance of an LF model.*

Despite the high value of hyper-parameters selection of an LFA model, there is relatively little research to address this issue. Current approaches can be divided into the following two categories:

(a) **Manually Tuning**. Manually tuning the hyper-parameters for an LFA model is a basic strategy. However, it is extremely hard for practitioners with little experience, and even non-trivial for experienced researchers. Therefore, it is not suitable for big data-based industrial applications, e.g., recommender system.

(b) **Hyper-parameters adaptation:** Most of the existing methods only make learning rate self-adaption in an LFA model, i.e., Adagrad [34], AdaDelta [35], and Adam [36]. However, they all have to fix the regularization coefficient. The choice of regularization coefficient is also very essential for the performance of an LFA model. Moreover, they all consume most time in computing the statistics of the past gradients, thereby cost much more time in a single iteration than a standard SGD algorithm does. Meanwhile, they cannot make the regularization coefficient self-adaptive either.

Hence, simultaneously adapting the learning rate and regularization coefficient for an LFA model in an efficient way is necessary. According to [37–40], a particle swarm optimization (PSO) algorithm is compatible with a multi-dimensional optimization problem with multiple decision parameters. From this point of view, we propose a *Learning Rate and Regularization Coefficient-free LFA* (LRLFA) model. Its main idea is to build a swarm by taking the learning rate and regularization coefficient of every single LFA-based model as particles, and then apply PSO to make them adaptation according to a pre-defined fitness function. Moreover, the linearly decreasing inertia weight is introduced to avoid PSO being stacked by local solutions. The main contributions of this chapter include:

(a) An LRLFA model. It can find the appropriate learning rate and regularization coefficient simultaneously.
(b) Algorithm design and analysis for an LRLFA model.

For validating LRLFA's performance, we have conducted empirical studies on six HiDS matrices. The positive experimental results verify the excellent performance of LRLFA.

The remainder of this chapter is organized as follows: Sect. 3.2 describes an SGD-based LFA model. Section 3.3 presents our LRLFA model in detail. Section 3.4 introduces the experimental results. Section 3.5 concludes the chapter and shows our future plan.

## 3.2  An SGD-Based LFA Model

As shown in Sect. 1.2.2, the objective function of an LFA model [41] with Euclidean distance is formulated by:

$$\varepsilon(X, Y) = \sum_{r_{m,n} \in \Lambda} \left( (r_{m,n} - \tilde{r}_{m,n})^2 + \lambda \|x_{m,.}\|_2^2 + \lambda \|y_{n,.}\|_2^2 \right)$$

$$= \sum_{r_{m,n} \in \Lambda} \left( (r_{m,n} - \tilde{r}_{m,n})^2 + \lambda \sum_{d=1}^{f} x_{m,d}^2 + \lambda \sum_{d=1}^{f} y_{n,d}^2 \right) \tag{3.1}$$

Where $\tilde{r}_{m,n} = \sum_{d=1}^{f} x_{m,d} y_{n,d}$, $\lambda$ denotes the regularization coefficient, $x_{m,}$ is the $m$-th row vector in $X$, $y_{n,}$ is the $n$-th row vector and $Y$, $r_{m,n}$, $x_{m,d}$ and $y_{n,d}$ denote the single elements of $R$, $X$ and $Y$, $\|\cdot\|_2$ calculates the $L_2$ norm of the enclosed vector.

As indicated by prior research [42], SGD is an efficient optimizer when building an LFA-based model of HiDS matrices. With it, the objective function (3.1) is minimized as follows:

$$\arg \min_{X,Y} \varepsilon(X, Y) \overset{SGD}{\Rightarrow} \forall r_{m,n} \in \Lambda, d \in \{1, 2, \ldots, f\} :$$

$$x_{m,d}^{\tau} \leftarrow x_{m,d}^{\tau-1} - \eta \frac{\partial \varepsilon_{m,n}^{\tau-1}}{\partial x_{m,d}^{\tau-1}}, y_{n,d}^{\sigma} \leftarrow y_{n,d}^{\sigma-1} - \eta \frac{\partial \varepsilon_{m,n}^{\sigma-1}}{\partial y_{n,d}^{\sigma-1}}; \qquad (3.2)$$

Where $\varepsilon_{m,n} = (r_{m,n} - \tilde{r}_{m,n})^2 + \lambda_X \|x_{m,.}\|_2^2 + \lambda_Y \|y_{n,.}\|_2^2$ is the instant error corresponding to the training instance $r_{m,n} \in \Lambda$, and $\tau$ and $(\tau - 1)$ denote the current and last update points for $x_{m,d}$, $\sigma$ and $\sigma - 1$ denote the current and last update points for $y_{n,d}$, respectively. Based on (3.1), we achieve the following expressions:

$$\frac{\partial \varepsilon_{m,n}^{\tau-1}}{\partial x_{m,d}^{\tau-1}} = - err_{m,d}^{\tau-1} \cdot y_{n,d}^{\sigma-1} + \lambda x_{m,d}^{\tau-1},$$

$$\frac{\partial \varepsilon_{m,n}^{\sigma-1}}{\partial y_{n,d}^{\sigma-1}} = - err_{m,d}^{\sigma-1} \cdot x_{m,d}^{\tau-1} + y_{n,d}^{\sigma-1}; \qquad (3.3)$$

where $err_{m,n} = r_{m,n} - \ddot{r}_{m,n}$, and substituting (3.3) into (3.2), we achieve the following update rules of LF matrices $X$ and $Y$ as follows:

$$\arg \min_{X,Y} \varepsilon(X, Y) \overset{SGD}{\Rightarrow} \forall r_{m,n} \in \Lambda, d \in \{1, 2, \ldots, f\} :$$

$$\begin{cases} x_{m,d}^{\tau} \leftarrow x_{m,d}^{\tau-1} + \eta \left( err_{m,d}^{\tau-1} \cdot y_{n,d}^{\sigma-1} - \lambda x_{m,d}^{\tau-1} \right), \\ y_{n,d}^{\sigma} \leftarrow y_{n,d}^{\sigma-1} + \eta \left( err_{m,d}^{\sigma-1} \cdot x_{m,d}^{\tau-1} - y_{n,d}^{\sigma-1} \right). \end{cases} \qquad (3.4)$$

As shown in (3.4), we can see that $\eta$ and $\lambda$ determine the performance of an LFA model. As supported by prior research [43, 44], the performance of such an LFA model really depends largely on the right choice of learning rate and regularization coefficient. Hence, how to find the appropriate values of these two hyper-parameters simultaneously is our concern.

## 3.3  The Proposed LRLFA Model

### 3.3.1  Learning Rate and Regularization Coefficient Adaptation via PSO

Note that we generate a swarm consists of $q$ particles, which fly in a 2-dimension space. The first dimension denotes learning rate and the second dimension denotes regularization coefficient. Therefore, in the $t$-th iteration, the velocity and position of each particle is as follows:

$$\begin{cases} v_j(t) = \left[v_{j\eta}(t), \ v_{j\lambda}(t)\right]^T \\ s_j(t) = \left[s_{j\eta}(t), \ s_{j\lambda}(t)\right]^T \end{cases} \tag{3.5}$$

where $v_{j\eta}$ and $v_{j\lambda}$ denote the velocity of the dimensional learning rate and regularization coefficient, $s_{j\eta}$ and $s_{j\lambda}$ indicate their velocity, respectively. Thus, we consider the problem as a two-dimensional vector optimization problem.

As shown in Sect. 1.2.3, the evolution rule regarding the $j$-th learning rate at the $t$-th iteration is:

$$\begin{cases} v_j(t+1) = \begin{bmatrix} v_{j\eta}(t+1) \\ v_{j\lambda}(t+1) \end{bmatrix} \\ \qquad = w \begin{bmatrix} v_{j\eta}(t) \\ v_{j\lambda}(t) \end{bmatrix} + c_1 r_1 \left( pb_j(t) - \begin{bmatrix} s_{j\eta}(t) \\ s_{j\lambda}(t) \end{bmatrix} \right) + c_2 r_2 \left( gb_i(t) - \begin{bmatrix} \eta_j(t) \\ \lambda_j(t) \end{bmatrix} \right), \\ s_j(t+1) = \begin{bmatrix} s_{j\eta}(t+1) \\ s_{j\lambda}(t+1) \end{bmatrix} = \begin{bmatrix} s_{j\eta}(t) \\ s_{j\lambda}(t) \end{bmatrix} + \begin{bmatrix} v_{j\eta}(t+1) \\ v_{j\lambda}(t+1) \end{bmatrix} \end{cases} \tag{3.6}$$

where $pb_j$ and $gb$ denotes the best particle location of particle and the best position of the whole swarm, $w$ is the non-negative inertia constant, $c_1$ and $c_2$ are cognitive and social coefficients, $r_1$ and $r_2$ are two uniform random numbers in the range of [0, 1], respectively. Note that in this chapter, only one hyper-parameter needs to be adaptation. Hence, the particles are searched in a one-dimensional space only.

Note that each particle is not permitted to move without boundary. Hence, we make a pre-defined range to prevent particles to fly out of the searching space. The pre-defined range is the hyper-parameters' candidate space in this chapter. For instance, the candidate space of learning rate is $[\eta_{min}, \eta_{max}]$, the candidate space of the regularization coefficient is $[\lambda_{min}, \lambda_{max}]$.

$$s_j(t+1) = \begin{bmatrix} s_{j\eta}(t+1) \\ s_{j\lambda}(t+1) \end{bmatrix} = \begin{cases} s_{j\eta}(t+1) = \begin{cases} \eta_{min}, & s_{j\eta}(t+1) \leq \eta_{min} \\ \eta_{max}, & s_{j\eta}(t+1) > \eta_{max} \end{cases} \\ s_{j\lambda}(t+1) = \begin{cases} \lambda_{min}, & s_{j\lambda}(t+1) \leq \lambda_{min} \\ \lambda_{max}, & s_{j\lambda}(t+1) > \lambda_{max} \end{cases} \end{cases} \tag{3.7}$$

Meanwhile, we also need to restrict the corresponding velocity of each particle to prevent particles to skip the optimal solution or fall into the local optimal solution.

$$v_j(t+1) = \begin{bmatrix} v_{j\eta}(t+1) \\ v_{j\lambda}(t+1) \end{bmatrix} = \begin{cases} v_{j\eta}(t+1) = \begin{cases} \hat{v}, & v_{j\eta}(t+1) \le \hat{v} \\ \check{v}, & v_{j\eta}(t+1) > \check{v} \end{cases} \\ v_{j\lambda}(t+1) = \begin{cases} \hat{v}, & v_{j\lambda}(t+1) \le \hat{v} \\ \check{v}, & v_{j\lambda}(t+1) > \check{v} \end{cases} \end{cases} \tag{3.8}$$

where $\hat{v}$ and $\check{v}$ mean the upper and lower bounds of $v$. In our study, we set the boundary of velocity as $\check{v} = 0.2 \times \min[(\eta_{max} - \eta_{min}), (\lambda_{max} - \lambda_{min})]$ [45] and $\hat{v} = -\check{v}$ [45], where $[\eta_{min}, \eta_{max}] = [2^{-12}, 2^{-8}]$ represents the boundary of the learning rate, and $[\lambda_{min}, \lambda_{max}] = [2^{-7}, 2^{-3}]$ denotes the boundary of the regularization coefficient, respectively.

Following the principle of PSO, we need to define the fitness function since it is crucial for PSO and determines the search direction. Note that this chapter adopts the missing data estimation of an HiDS matrix as the representation learning objective. When addressing such a task, prediction accuracy for missing values is the key performance indicator, which is commonly measured by root mean squared error (RMSE) and mean absolute error (MAE). Therefore, we adopt RMSE as the fitness functions:

$$F_1(s) = \sqrt{\left( \sum_{r_{m,n} \in \Omega} (r_{m,n} - \tilde{r}_{m,n})^2 \right) / |\Omega|},$$

$$F_2(s) = \left( \sum_{r_{m,n} \in \Omega} |r_{m,n} - \tilde{r}_{m,n}|_{abs} \right) / |\Omega|; \tag{3.9}$$

where $\Omega$ denotes the validating set and is disjoint with $\Lambda$ and $\Phi$, $|\cdot|_{abs}$ calculate the absolute value of given value, $\tilde{r}_{m,n}$ denotes the prediction value corresponding to the instance $r_{m,n} \in \Omega$. Naturally, the fitness function should be reconsidered when the optimization objective of the whole swarm alters.

Based on (3.4)–(3.9), the update rules of LF matrices $X$ and $Y$ are showing as:

$$\underset{X,Y}{\arg\min}\ \varepsilon(X, Y) \overset{SGD}{\Rightarrow} \forall r_{m,n} \in \Lambda, d \in \{1, 2, \ldots, f\} :$$

$$\begin{cases} x_{m,d}^{\tau} \leftarrow x_{m,d}^{\tau-1} - s_{j\eta}(t)\left(s_{j\lambda}(t)x_{m,d}^{\tau-1} - err_{m,d}^{\tau-1} \cdot y_{n,d}^{\sigma-1}\right), \\ y_{n,d}^{\sigma} \leftarrow y_{n,d}^{\sigma-1} - s_{j\eta}(t)\left(s_{j\lambda}(t)y_{n,d}^{\sigma-1} - err_{m,d}^{\sigma-1} \cdot x_{m,d}^{\tau-1}\right). \end{cases} \tag{3.10}$$

From (3.10) and Fig. 3.2, we can see that the corresponding LF matrices of each particle (learning rate and regularization coefficient) is updated by corresponding matrices of the previous particle. In our scene, the particle is learning rate and regularization coefficient instead of LFs. Therefore, if we set separate matrices for

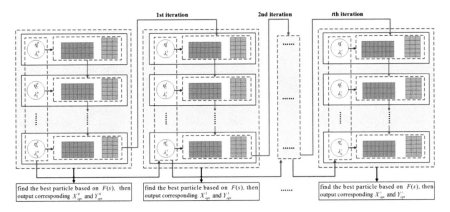

**Fig. 3.2** Structure and processing flow of LRLFA

each particle, it can not make up an efficient swarm due to the weak connections between particles.

### 3.3.2 Linearly Decreasing Inertia Weight Incorporation

A standard PSO has the advantages of high efficiency and good compatibility [37–39]. Nevertheless, it is easy to suffer from premature convergence and fall into local solutions. Therefore, linearly decreasing inertia weight [32] is introduced into the evolutionary process to further improve the accuracy of LRLFA. The inertia weight $w$ varies linearly with the generations, which is given by:

$$w^t = w_{max} - (w_{max} - w_{min}) \times \frac{t_{round}}{t_{max}} \tag{3.11}$$

where $w_{max}$ and $w_{min}$ are the maximal and minimal value of the inertia weight, which are 0.9 and 0.4 respectively according to [45]. $t_{round}$ represents the current iteration number and $t_{max}$ represents the maximum number of training iterations.

Generally, reasonable control of the two acceleration coefficients helps improve the performance of PSO. Therefore, we make two acceleration coefficients change with time to reduce the local optima component and increase the global component as follows:

$$c_1^t = c_{max} - (c_{max} - c_{min}) \times \frac{t_{round}}{t_{max}}$$

$$c_2^t = c_{min} + (c_{max} - c_{min}) \times \frac{t_{round}}{t_{max}} \tag{3.12}$$

where $c_{max}$ and $c_{min}$ represent the maximal and minimal values of two acceleration coefficients, where their values are set as 2.5 and 0.5 according to [46], respectively.

### 3.3.3   Algorithm Design and Analysis

Based on the above analysis, we develop an LRLFA Algorithm. In each iteration, LRLFA updates LF matrices $X$ and $Y$ with (3.10), and updates the hyper-parameters with (3.6). As illustrated in Algorithm LRLFA, $v$ and $s$ represent the velocity and position vector of the particle, $pb$ and $gb$ represent the local optimal position and the global optimal position, and $z$ denotes the value of the fitness function $F$, respectively.

| Procedure WeightAdjust | |
| --- | --- |
| **Input: $E, n$** | |
| **Operation** | **Cost** |
| $w = w_{max} - (w_{max} - w_{min}) \times (n/E)$ | $\Theta(1)$ |
| $C_1 = C_{max} - (C_{max} - C_{min}) \times (n/E)$ | $\Theta(1)$ |
| $C_2 = C_{min} + (C_{max} - C_{min}) \times (n/E)$ | $\Theta(1)$ |
| **Output: $w, c_1, c_2$** | |

As shown in Algorithm LRLFA, we calculate its time complexity as follows:

$$T_{HLFA} \approx \Theta\big(((|M| + |N|) \times f + 3q \times 2 + 2 + q) + \Theta(n \times q \times (|\Lambda| \times 3f + |\Omega| \times f))$$
$$\approx \Theta(n \times q \times |\Lambda| \times f) \tag{3.13}$$

Note that in (3.13) with $|\Lambda| \ll \max\{|M|, |N|\}$, we can omit the lower-order-terms to ensure the result reasonably. There are two main factors when computing the computational complexity of LRLFA: (1) An SGD-based LF model's matrices $X$ and $Y$, whose storage cost is $\Theta((|M| + |N|) \times f)$; (2) the particle of PSO's auxiliary matrices $S, V, pb$ whose storage cost is:

$$S_{HLFA} \approx (|M| + |N|) \times f + 6q \tag{3.14}$$

Based on the above inferences, we see that the LRLFA algorithm is highly efficient in both computation and storage.

| Algorithm LRLFA | |
| --- | --- |
| **Input: $U, I, \Lambda, f, \Gamma$** | |
| **Operation** | **Cost** |
| **Initialize $X^{|M| \times f}$ and $Y^{|N| \times f}$ randomly** | $\Theta((|M| + |N|) \times f)$ |
| **Initialize $q, r_1, r_2, c_1, c_2, w, \eta_{min}, \eta_{max}, \lambda_{min}, \lambda_{max}, v_{initmin}, v_{initmax}, w_{max}, w_{min},$** $c_{max}, c_{min}$ | $\Theta(1)$ |

(continued)

| **Algorithm LRLFA** | |
|---|---|
| **Initialize** $S^{q \times 2}$, $V^{q \times 2}$ and $pb^{q \times 2}$ randomly | $\Theta(3q \times 2)$ |
| **Initialize** $gb_2$ and $z_q$ randomly | $\Theta(2 + q)$ |
| **Initialize** $n = 0$, *Max-training-round* $= E$ | $\Theta(1)$ |
| **while not** converge **and** $n \leq E$ **do** | $\times n$ |
|   **for** $j = 1$ **to** $q$ | $\times q$ |
|     **for each** $r_{m,n} \in \Lambda$ | $\times |\Lambda|$ |
|     $err_{m,n} = r_{m,n} - \sum_{d=1}^{f} x_{m,d} y_{n,d}$ | $\Theta(f)$ |
|     **for** $d = 1$ **to** $f$ | $\times f$ |
|     $x_{m,d} = x_{m,d} - s_{j\eta}(s_{j\lambda}x_{m,d} - y_{n,d}err_{m,n}), y_{n,d} = y_{n,d} - s_{j\eta}(s_{j\lambda}y_{n,d} - x_{m,d}err_{m,n})$ | $\Theta(1)$ |
|     **end for** | – |
|     **end for** | – |
|   $z_j = F(s_j)$ according to (3.9) | $\Theta(|\Omega| \times f)$ |
|   **end for** | – |
|   **for** $j = 1$ **to** $q$ | $\times q$ |
|     **if** $z_j < F(pb_j)$ **then** $pb_j = s_j$; **if** $z_j < F(gb)$ **then** $gb = s_j$ | $\Theta(1)$ |
|   **end for** | – |
|   **for** $j = 1$ **to** $q$ | $\times q$ |
|     $(w, c_1, c_2) = $ **WeightAdjust**$(E, n)$ | $\Theta(3)$ |
|     $v_j = wv_j + c_1 r_1(pb_j - s_j) + c_2 r_2(gb - s_j)$ | $\Theta(1)$ |
|     **if** $v_j < = \breve{v}$ **then** $v_j = \breve{v}$ | $\Theta(1)$ |
|     **else if** $v_j > v\hat{}$ **then** $v_j = v\hat{}$ | $\Theta(1)$ |
|     $s_j = s_j + v_j$ | $\Theta(2)$ |
|     **if** $\eta_j < = \eta_{min}$ **then** $\eta_j = \eta_{min}$ | $\Theta(1)$ |
|     **else if** $\eta_j > \eta_{max}$ **then** $\eta_j = \eta_{max}$ | $\Theta(1)$ |
|     **if** $\lambda_j < = \lambda_{min}$ **then** $\lambda_j = \lambda_{min}$ | $\Theta(1)$ |
|     **else if** $\lambda_j > \lambda_{max}$ **then** $\lambda_j = \lambda_{max}$ | $\Theta(1)$ |
|   **end for** | – |
|   $n = n + 1$ | $\Theta(1)$ |
| **end while** | – |
| **Output:** $X$, $Y$ | |

## 3.4  Experimental Results and Analysis

### 3.4.1  General Settings

**Evaluation Metrics** For industrial applications, one major motivation to processing an HiDS matrix is to predict its missing values. We adopt it as the evaluation protocol in our experiments owing to its popularity and usefulness. It is commonly measured by root mean squared error (RMSE) and mean absolute error (MAE) [47–50]:

**Table 3.1** Datasets details

| No. | Name | Row | Column | Known Entries | Density (%) |
|-----|------|-----|--------|---------------|-------------|
| D1 | MovieLens 10M | 10,681 | 71,567 | 10,000,054 | 1.31 |
| D2 | Douban | 58,541 | 129,490 | 16,830,839 | 0.22 |
| D3 | EachMovie | 72,916 | 1628 | 2,811,718 | 2.37 |
| D4 | Flixter | 48,794 | 147,612 | 8,196,077 | 0.11 |
| D5 | Epinion | 120,492 | 775,760 | 13,668,320 | 0.015 |
| D6 | MovieLens 20M | 26,744 | 138,493 | 20,000,263 | 0.54 |

$$RMSE = \sqrt{\left(\sum_{r_{m,n}\in\Phi}(r_{m,n} - \tilde{r}_{m,n})^2\right)/|\Phi|,}$$

$$RMSE = \left(\sum_{r_{m,n}\in\Phi}|r_{m,n} - \tilde{r}_{m,n}|_{abs}\right)/|\Phi|;$$

where $\ddot{r}_{m,n}$ denotes the generated prediction for the testing instance $r_{m,n}\in\Phi$, which represents the missing value at the $m$-th row and $n$-th column of a target HiDS matrix. $|\cdot|$ represents the cardinality of a given set. $|\Phi|$ denotes the size of the validation dataset $\Omega$.

**Datasets**  Six HiDS matrices are included in our experiment. In order to verify the feasibility of the LRLFA model in addressing big data from real applications, all datasets are real datasets collected by industrial companies, the datasets details are recorded in Table 3.1.

Note that each dataset is randomly split into ten disjoint subsets for implementing 70%–10%–20% train-validation-test settings. More specifically, on each dataset, we adopt seven subsets as a training set to train a model, one as a validating set to monitor the training process for making the model achieve its optimal outputs, and the last two as the testing set for verifying the performance of each tested model. This process is sequentially repeated ten times to acquire the final results. The termination condition is uniform for all involved models, i.e., the iteration threshold is 1000, and error threshold is $10^{-5}$, and a tested model's training process terminates if either threshold is met.

**Model Settings**  To obtain a fair and objective comparison, we adopt the following measures settings:

(a) LF matrices $X$ and $Y$ are initialized randomly generated for eliminating the performance bias; and
(b) LF dimension $f = 20$ according to prior researches [28, 45] to balance the representative learning ability and computational efficiency of each model.
(c) For PSO, the dimension space $D = 2$, the search space of the learning rate is [$2^{-12}$, $2^{-8}$], and the search space of the regularization coefficient is [$2^{-7}$, $2^{-3}$]. $r_1$

and $r_2$ are two uniformly distributed random numbers in a range of [0, 1] according to [45].

## 3.4.2 Effect of Swarm Size

The swarm size affects LRLFA's performance. Figures 3.3, 3.4, and 3.5 show the swarm size's effects in the prediction accuracy and time cost per iteration of LRLFA. From them, the following findings are achieved:

(a) LRLFA's prediction accuracy is affected by the swarm size. As depicted in Figs. 3.3 and 3.4, LRLFA's prediction accuracy increases as the swarm size increases. For instance, on D1, the lowest RMSE of LRLFA is 0.7916, 0.787, 0.7866, 0.7852, 0.7854 and 0.7851 when $q = 5$, 10, 20, 40, 60, and 80, respectively. The lowest RMSE is 0.7851 with $q = 80$ and the highest RMSE is 0.7916 with $q = 5$, the gap between $q = 5$ and $q = 80$ reaches 0.82%. However, the situation is different on D3 and D6. When swarm size surpasses a certain threshold, the prediction accuracy decreases. This phenomenon is reasonable since the selection of swarm size is related to the actual problem.

(b) LRLFA's time cost per iteration increases linearly as swarm size increases. For instance, as depicted in Fig. 3.5a, on D1, for the case of RMSE, LRLFA's time cost per iteration is 2.90, 5.89, 11.73, 23.40, 35.69 and 46.65 s as $q = 5$, 10, 20, 40, 60 and 80, respectively. Similar outcomes are found in other testing cases, which is consistent with LRLFA's theoretical computational complexity.

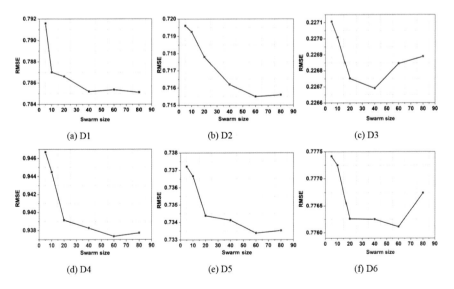

**Fig. 3.3** LRLFA's RMSE as swarm size varies

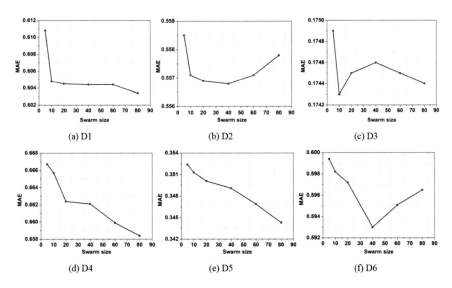

**Fig. 3.4** LRLFA's MAE as swarm size varies

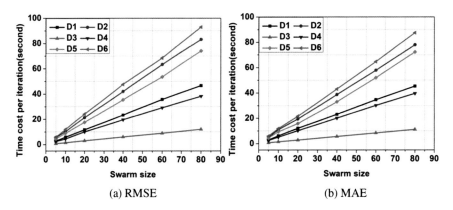

**Fig. 3.5** LRLFA's time cost per iteration as swarm size varies

Based on these results mentioned above, it is necessary to pick up appropriate swarm size. Note that swarm size is more sensitive to computational efficiency in our context. Therefore, with a comprehensive consideration of prediction accuracy and computational efficiency, we fix $q = 10$ in the following experiments.

### 3.4.3  Performance Comparison

In this part of experiments, we compare the proposed LRLFA with the most widely used state-of-the-art methods. The details of compared models are summarized as follows:

(a) M1: an SGD-based LF model [42]. For this original LF model, we select the appropriate learning rate and regularization coefficient using a grid-search-method.
(b) M2: An RMSprop-based LFA model [42]. It replaces the sum squares of the past stochastic gradients with their decaying average to make its learning rate self-adaptive.
(c) M3: An Ada-Delta-based LFA model [42]. Similar to M2, it also considers the past stochastic gradients at previous update points. However, it controls the historical learning information more carefully by recording its decaying square average of the past update increments.
(d) M4: an Adam-based LFA model [42], which adjusts the learning rate based on both the exponentially decaying square average and the exponentially decaying average of past stochastic gradients.
(e) M5: An LRLFA model proposed in this study.
(f) M6: An LRLFA model incorporating linearly decreasing inertia weight.

The lowest RMSE and MAE are shown in Figs. 3.6 and 3.7, Figs. 3.8 and 3.9 depict the training curves of compared models on validation set, the time cost per iteration and iteration of M1–M6 are summarized in Tables 3.2 and 3.3, the total time cost of are shown in Figs. 3.10 and 3.11. From them, we have the following finding:

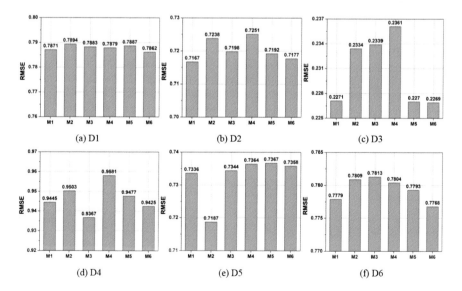

**Fig. 3.6** Lowest RMSE of M1–M6

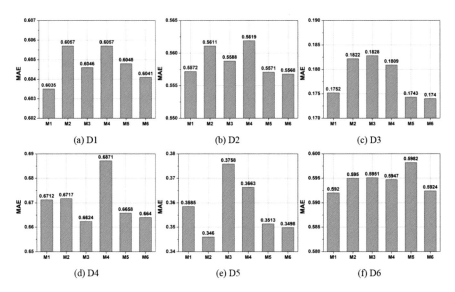

**Fig. 3.7**  Lowest MAE of M1–M6

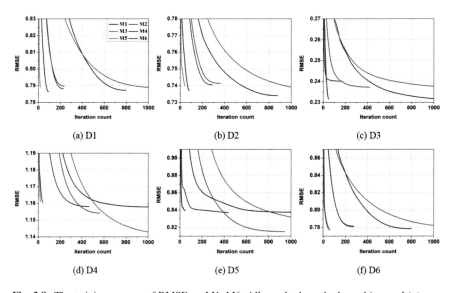

**Fig. 3.8**  The training process of RMSE on M1–M6. All panels share the legend in panel (**a**)

(a) LRLFA achieves competitive prediction accuracy for missing data of an HiDS matrix when compared with its peers. For instance, as shown in Fig. 3.4, on D1, D3, D4 and D6, LRLFA's lowest RMSE are 0.7862, 0.2269, 0.9425 and 0.7768, about 0.11%, 0.09%, 0.2%, and 0.14% lower than that of M1, respectively. However, on D2 and D5, the RMSE is a little higher than that of M1's about 0.14% and 0.3%. Similar outcomes also exist on MAE.

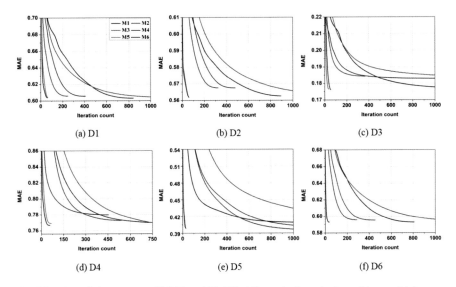

**Fig. 3.9** The training process of MAE on M1–M6. All panels share the legend in panel (**a**)

**Table 3.2** Time cost per iteration (seconds) and iterations of RMSE

| Dataset | Time cost per iteration/Iterations | | | | | |
|---|---|---|---|---|---|---|
| | M1 | M2 | M3 | M4 | M5 | M6 |
| D1 | 0.576/798 | 1.781/235 | 3.176/998 | 3.397/228 | 6.667/**18** | 6.359/92 |
| D2 | 1.076/877 | 3.206/365 | 5.631/1000 | 5.523/291 | 10.268/**41** | 10.488/82 |
| D3 | 0.153/1000 | 0.528/423 | 0.846/1000 | 0.879/167 | 1.510/**51** | 1.922/51 |
| D4 | 0.460/1000 | 1.449/562 | 2.666/1000 | 2.774/466 | 4.947/**19** | 4.682/44 |
| D5 | 0.913/1000 | 2.964/945 | 4.855/1000 | 4.627/438 | 8.902/**41** | 8.745/47 |
| D6 | 1.266/799 | 3.919/276 | 6.163/1000 | 7.010/271 | 12.04/**50** | 12.328/61 |

**Table 3.3** Time cost per iteration (seconds) and iteration count of MAE

| Dataset | Time cost per iteration/Iterations | | | | | |
|---|---|---|---|---|---|---|
| | M1 | M2 | M3 | M4 | M5 | M6 |
| D1 | 0.640/841 | 1.988/248 | 3.208/1000 | 2.977/399 | 6.164/56 | 6.513/**51** |
| D2 | 1.135/892 | 3.648/324 | 5.519/1000 | 5.242/479 | 10.971/**57** | 10.022/61 |
| D3 | 0.178/998 | 0.544/367 | 0.819/1000 | 0.798/1000 | 1.487/**37** | 1.49/55 |
| D4 | 0.543/936 | 1.637/539 | 2.609/1000 | 3.033/453 | 5.109/68 | 4.767/**58** |
| D5 | 1.032/998 | 3.007/1000 | 4.612/1000 | 4.197/1000 | 9.353/**23** | 8.985/29 |
| D6 | 1.395/810 | 4.074/289 | 6.211/1000 | 6.035/453 | 11.868/**16** | 11.341/37 |

(b) LRLFA's convergence rate is fast. For instance, as shown in Table 3.2 and Fig. 3.6, M1 takes 798, 877, 1000, 1000, 1000, 799 iterations to get the lowest RMSE on D1–D6, respectively. While M5, i.e., the proposed LRLFA model, takes about 18, 41, 51, 19, 41, and 50 iterations to reach the lowest RMSE on

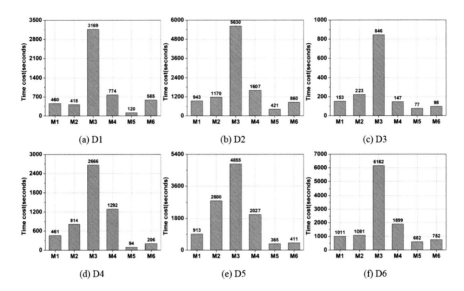

**Fig. 3.10**  Total time cost of RMSE on M1–M6

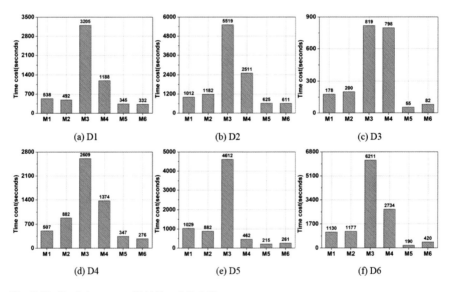

**Fig. 3.11**  Total time cost of MAE on M1–M6

D1–D6, respectively. It reduces 97.74%, 95.32%, 94.9%, 98.1%, 95.9% and 93.74% iterations than that of M1. Similar situations are found on MAE.

(c) LRLFA's computational efficiency is high. As depicted in Fig. 3.8, on D1, M5 consumes 120 s to achieve its lowest RMSE. This is 73.91% of 460 s by M1, 71.29% of 418 s by M2, 96.21% of 3169 s by M3, 84.50% of 774 s by M4, and

79.49% of 585 s by M6. However, on D1, D2 and D4, for the case of MAE, M6's computational efficiency is higher than that of M5's Remarkably, M6's total time cost is slightly higher than that of M5.

### 3.4.4 Summary

Based on the experimental results, we summarize that:

(a) The swarm size should be carefully selected with a comprehensive consideration of prediction accuracy and computational efficiency for the proposed LRLFA model;
(b) LRLFA can adjust the learning rate and regularization coefficient simultaneously in one full training process;
(c) LRLFA greatly reduces computational cost needed to search for appropriate learning rate and regularization coefficient.
(d) LRLFA achieves competitive prediction accuracy and high computational efficiency for missing data of an HiDS matrix when compared with the state-of-the-art methods

## 3.5 Conclusions

Selecting the appropriate learning rate and regularization coefficient is tedious but extremely important for an LFA model. To address this problem, we innovatively propose an LRLFA model that implements efficient and self-adaptive selection of the optimal learning rate and regularization coefficient. Empirical studies on public HiDS matrices demonstrate that LRLFA avoids a time-consuming search for these two hyper-parameters. Furthermore, it also achieves highly-competitive prediction accuracy for the missing data of an HiDS matrix.

In future, we aim to integrate more advanced PSO algorithms [51, 52] into an LRLFA model to improve its both prediction accuracy and time efficiency. Moreover, we plan to commit to applying this optimization scheme to a broader field, such as robot calibration and control [53–60], Network representation [61] and the applications in other fields [62], thereby achieving more promising applications.

## References

1. Gatzioura, A., Sanchez-Marre, M.: A case-based recommendation approach for market basket data. IEEE Intell. Syst. **30**(1), 20–27 (2015)

2. Adomavicius, G., Tuzhilin, A.: Toward the next generation of recommender systems: a survey of the state-of-the-art and possible extensions. IEEE Trans. Knowl. Data Eng. **17**(6), 734–749 (2005)
3. Luo, X., Zhou, M.C., Li, S., Xia, Y.N., You, Z.H., Zhu, Q.S., Leung, H.: An efficient second-order approach to factorizing sparse matrices in recommender systems. IEEE Trans. Ind. Inform. **11**(4), 946–956 (2015)
4. Sarwar, B., Karypis, G., Konstan, J., Reidl, J.: Item-based collaborative-filtering recommendation algorithms. In: Proc. of the 10th Int. Conf. World Wide Web, pp. 285–295 (2001)
5. Zhang, J.B., Lin, Z.Q., Xiao, B., Zhang, C.: An optimized item-based collaborative filtering recommendation algorithm. In: IEEE Int. Conf. on Network Infrastructure and Digital Content, pp. 414–418 (2009)
6. Wu, D., He, Y., Luo, X., Zhou, M.C.: A latent factor analysis-based approach to online sparse streaming feature selection. IEEE Trans. Syst. Man Cybern. Syst., 1–15. https://doi.org/10.1109/TSMC.2021.3096065
7. Wang, Z., Liu, Y., Luo, X., Wang, J.J., Gao, C., Peng, D.Z., Chen, W.: Large-scale affine matrix rank minimization with a novel nonconvex regularizer. IEEE Trans. Neural Netw. Learn. Syst. **33**(9), 4661–4675 (2022). https://doi.org/10.1109/TNNLS.2021.3059711
8. Andersen, R., Borgs, C., Chayes, J., Feige, U., Flaxman, A., Kalai, A., Mirrokni, V., Tennenholtz, M.: Trust-based recommendation systems: an axiomatic approach. In: Proc. of the 17th Int. Conf. on World Wide Web ACM, pp. 199–208 (Apr 2008)
9. Li, Y.F., Ngom, A.: Versatile sparse matrix factorization and its applications in high-dimensional biological data analysis. In: Proc of the 8th Int. Conf. on Pattern Recognition in Bioinformatics Springer, pp. 91–101 (2013)
10. Yuan, Y., He, Q., Luo, X., Shang, M.S.: A multilayered-and-randomized latent factor model for high-dimensional and sparse matrices. IEEE Trans. Big Data. **8**(3), 784–794 (2022). https://doi.org/10.1109/TBDATA.2020.2988778
11. Wu, H., Luo, X., Zhou, M.C., Rawa, M.J., Sedraoui, K., Albeshri, A.: A PID-incorporated latent factorization of tensors approach to dynamically weighted directed network analysis. IEEE/CAA J. Autom. Sin. **9**(3), 533–546 (2022). https://doi.org/10.1109/JAS.2021.1004308
12. Luo, X., Wang, Z.D., Shang, M.S.: An instance-frequency-weighted regularization scheme for non-negative latent factor analysis on high-dimensional and sparse data. IEEE Trans. Syst. Man Cybern. Syst. **51**(6), 3522–3532 (2021)
13. Luo, X., Qin, W., Dong, A., Sedraoui, K., Zhou, M.C.: Efficient and high-quality recommendations via momentum-incorporated parallel stochastic gradient descent-based learning. IEEE/CAA J. Autom. Sin. **8**(2), 402–411 (2021)
14. Wu, D., Luo, X., Shang, M., He, Y., Wang, G.Y., Zhou, M.C.: A deep latent factor model for high-dimensional and sparse matrices in recommender systems. IEEE Trans. Syst. Man Cybern. Syst. **51**(7), 4285–4296 (2021)
15. Wu, D., Luo, X.: Robust latent factor analysis for precise representation of high-dimensional and sparse data. IEEE/CAA J. Autom. Sin. **8**(4), 796–805 (2021)
16. Luo, X., Zhou, Y., Liu, Z.G., Zhou, M.C.: Fast and accurate non-negative latent factor analysis on high-dimensional and sparse matrices in recommender systems. In: IEEE Trans. Knowl. Data Eng. https://doi.org/10.1109/TKDE.2021.3125252
17. Luo, X., Zhong, Y.R., Wang, Z.D., Li, M.Z.: An alternating-direction-method of multipliers-incorporated approach to symmetric non-negative latent factor analysis. In: IEEE Trans. Neural Netw. Learn. Syst. https://doi.org/10.1109/TNNLS.2021.3125774
18. Luo, X., Wu, H., Li, Z.C.: NeuLFT: a novel approach to nonlinear canonical polyadic decomposition on high-dimensional incomplete tensors. IEEE Trans. Knowl. Data Eng. https://doi.org/10.1109/TKDE.2021.3176466
19. Jing, Y.C., Zhang, X.Z., Wu, L.F., Wang, J.Q., Feng, Z.M., Wang, D.: Recommendation on Flickr by combining community user ratings and item importance. In Proc. of Inte. Conf. on Multimedia and Expo (ICME), pp. 1–6 (2012)

20. Ma, H., King, I., Lyu, M.R.: Learning to recommend with social trust ensemble. In: Proc. of the 32nd Int. ACM SIGIR Conf. on Research and Development in Information Retrieval, pp. 203–210 (2009)
21. Luo, X., Zhou, M., Li, S., Shang, M.: An inherently non-negative latent factor model for high-dimensional and sparse matrices from industrial applications. IEEE Trans. Ind. Inform. 14(5), 2011–2022 (2018)
22. Qin, W.J., Wang, H.L., Zhang, F., Wang, J.J., Luo, X., Huang, T.W.: Low-rank high-order tensor completion with applications in visual data. IEEE Trans. Image Process. 31, 2433–2448 (2022). https://doi.org/10.1109/TIP.2022.3155949
23. Song, Y., Zhu, Z.Y., Li, M., Yang, G.S., Luo, X.: Non-negative latent factor analysis-incorporated and feature-weighted fuzzy double c-means clustering for incomplete data. IEEE Trans. Fuzzy Syst. https://doi.org/10.1109/TFUZZ.2022.3144489
24. Li, W.L., He, Q., Luo, X., Wang, Z.D.: Assimilating second-order information for building non-negative latent factor analysis-based recommenders. IEEE Trans. Syst. Man Cybern. Syst. 52(1), 485–497 (2021)
25. Luo, X., Yuan, Y., Zhou, M.C., Liu, Z.G., Shang, M.S.: Non-negative latent factor model based on β-divergence for recommender systems. IEEE Trans. Syst. Man Cybern. Syst. 51(8), 4612–4623 (2021)
26. Luo, X., Zhou, M.C., Li, S., Wu, D., Liu, Z.G., Shang, M.S.: Algorithms of unconstrained non-negative latent factor analysis for recommender systems. IEEE Trans. Big Data. 7(1), 227–240 (2021)
27. Liu, Z.G., Luo, X., Wang, Z.D.: Convergence analysis of single latent factor-dependent, non-negative and multiplicative update-based non-negative latent factor models. IEEE Trans. Neural Netw. Learn. Syst. 32(4), 1737–1749 (2021)
28. Wijnhoven, R.G.J., de With, P.H.N.: Fast training of object detection using stochastic gradient descent. In: 2010 20th Int. Conf on Pattern Recognition. https://doi.org/10.1109/ICPR.2010.112
29. Xie, X., et al.: CuMF_SGD: parallelized stochastic gradient descent for matrix factorization on GPUs. In: ACM International Symposium on High-Performance Parallel and Distributed Computing (HPDC). https://doi.org/10.1145/3078597.3078602
30. Chen, L., Wang, J.: Dictionary learning with weighted stochastic gradient descent. In: 2012 Int. Conf. on Computational Problem-Solving (ICCP). https://doi.org/10.1109/ICCPS.2012.6384229
31. Watanabe, T., Iima, H.: Nonlinear optimization method based on stochastic gradient descent for fast convergence. In: 2018 IEEE Int. Conf. on Systems, Man, and Cybernetics (SMC). https://doi.org/10.1109/SMC.2018.00711
32. Li, S., You, Z.H., Guo, H.L., Luo, X., Zhao, Z.Q.: Inverse-free extreme learning machine with optimal information updating. IEEE Trans. Cybern. 46(5), 1229–1241 (2016)
33. Luo, X., Zhou, M.C., Xia, Y.N., Zhu, Q.S.: An efficient non-negative matrix-factorization-based approach to collaborative filtering for recommender systems. IEEE Trans. Ind. Inform. 10(2), 1273–1284 (2014)
34. Duchi, J., Hazan, E., Singer, Y.: Adaptive subgradient methods for online learning and stochastic optimization. J. Mach. Learn. Res. 12(7), 2121–2159 (2011)
35. Zeiler, M.D.: ADADELTA: an adaptive learning rate method. Comput. Sci. (2012)
36. Kingma, D., Ba, J.: Adam: a method for stochastic optimization. In Int. Conf. for Learning Representations (2015)
37. Shi, Y., Eberhart, R.C.: Empirical study of particle swarm optimization. In: Proc. of the 1999 IEEE Congress on Evolutionary Computation, pp. 1945–1950 (1999)
38. Chenga, R., Jin, Y.C.: A social learning particle swarm optimization algorithm for scalable optimization. Inf. Sci. 291, 43–60 (2015)
39. Cao, Y.L., Zhang, H., Li, W.F., Zhou, M.C., Zhang, Y., Chaovalitwongse, W.A.: Comprehensive learning particle swarm optimization algorithm with local search for multimodal functions. IEEE Trans. Evol. Comput. 23(4) (2019)

40. Luo, X., Yuan, Y., Chen, S.L., Zeng, N.Y., Wang, Z.D.: Position-transitional particle swarm optimization-incorporated latent factor analysis. IEEE Trans. Knowl. Data Eng. **34**(8), 3958–3970 (2022). https://doi.org/10.1109/TKDE.2020.3033324

41. Wu, D., Shang, M.S., Luo, X., Wang, Z.D.: An $L_1$-and-$L_2$-norm-oriented latent factor model for recommender systems. IEEE Trans. Neural Netw. Learn. Syst. https://doi.org/10.1109/TNNLS. 2021.3071392

42. Luo, X., Wang, D.X., Zhou, M.C., Yuan, H.Q.: Latent factor-based recommenders relying on extended stochastic gradient descent algorithms. IEEE Trans. Syst. Man Cybern. Syst. **51**(2), 916–926 (2021)

43. Shang, F., Jiao, L.C., Wang, F.: Graph dual regularization non-negative matrix factorization for co-clustering. Pattern Recogn. **45**(6), 2237–2250 (2012)

44. Hu, L., Yan, S.C., Luo, X., Zhou, M.C.: An algorithm of inductively identifying clusters from attributed graphs. IEEE Trans. Big Data. **8**(2), 523–534 (2022). https://doi.org/10.1109/ TBDATA.2020.2964544

45. Eberhart, R.C., Shi, Y.H.: Particle swarm optimization: developments, applications and resources. In: Proc. of the Congress on Evolutionary Computation, pp. 81–86 (2001)

46. Ratnaweera, A., Halgamuge, S.K., Watson, H.C.: Self-organizing hierarchical particle swarm optimizer with time-varying acceleration coefficients. IEEE Trans. Evol. Comput. **8**(3), 240–255 (2004)

47. Luo, X., Wu, H., Yuan, H.Q., Zhou, M.C.: Temporal pattern-aware QoS prediction via biased non-negative latent factorization of tensors. IEEE Trans. Cybern. **50**(5), 1798–1809 (2020)

48. Song, Y., Li, M., Luo, X., Yang, G.S., Wang, C.J.: Improved symmetric and nonnegative matrix factorization models for undirected, sparse and large-scaled networks: a triple factorization-based approach. IEEE Trans. Ind. Inform. **16**(5), 3006–3017 (2020)

49. Shi, X.Y., He, Q., Luo, X., Bai, Y.N., Shang, M.S.: Large-scale and scalable latent factor analysis via distributed alternative stochastic gradient descent for recommender systems. IEEE Trans. Big Data. **8**(2), 420–431 (2022). https://doi.org/10.1109/TBDATA.2020.2973141

50. Shang, M.S., Yuan, Y., Luo, X., Zhou, M.C.: An $A$-$\beta$-$divergence$-generalized recommender for highly-accurate predictions of missing user preferences. IEEE Trans. Cybern. **52**(8), 8006–8018 (2022). https://doi.org/10.1109/TCYB.2020.3026425

51. Jiang, F., et al.: A new binary hybrid particle swarm optimization with wavelet mutation. Knowl.-Based Syst. **130**, 90–101 (2017)

52. Wu, X.: A density adjustment based particle swarm optimization learning algorithm for neural network design. In: 2011 Int. Conf. on Electrical and Control Engineering. https://doi.org/10. 1109/ICECENG.2011.6057937

53. Li, Z.B., Li, S., Bamasag, O., Alhothali, A., Luo, X.: Diversified regularization enhanced training for effective manipulator calibration. IEEE Trans. Neural Netw. Learn. Syst. https:// doi.org/10.1109/TNNLS.2022.3153039

54. Zhang, F., Jin, L., Luo, X.: Error-summation enhanced newton algorithm for model predictive control of redundant manipulators. IEEE Trans. Ind. Electron. https://doi.org/10.1109/TIE. 2022.3165277

55. Li, Z.B., Li, S., Luo, X.: An overview of calibration technology of industrial robots. IEEE/CAA J. Autom. Sin. **8**(1), 23–36 (2021)

56. Khan, A.H., Li, S., Luo, X.: Obstacle avoidance and tracking control of redundant robotic manipulator: an RNN based metaheuristic approach. IEEE Trans. Ind. Inform. **16**(7), 4670–4680 (2020)

57. Chen, D.C., Li, S., Wu, Q., Luo, X.: New disturbance rejection constraint for redundant robot manipulators: an optimization perspective. IEEE Trans. Ind. Inform. **16**(4), 2221–2232 (2020)

58. Xie, Z.T., Jin, L., Luo, X., Sun, Z.B., Liu, M.: RNN for repetitive motion generation of redundant robot manipulators: an orthogonal projection based scheme. IEEE Trans. Neural Netw. Learn. Syst. **33**(2), 615–628 (2022). https://doi.org/10.1109/TNNLS.2020.3028304

59. Xie, Z.T., Jin, L., Luo, X., Li, S., Xiao, X.C.: A data-driven cyclic-motion generation scheme for kinematic control of redundant manipulators. IEEE Trans. Control Syst. Technol. **29**(1), 53–63 (2021)
60. Jin, L., Qi, Y.M., Luo, X., Li, S., Shang, M.S.: Distributed competition of multi-robot coordination under variable and switching topologies. IEEE Trans. Autom. Sci. Eng. https://doi.org/10.1109/TASE.2021.3126385
61. Chen, X.F., Luo, X., Jin, L., Li, S., Liu, M.: Growing echo state network with an inverse-free weight update strategy. IEEE Trans. Cybern. https://doi.org/10.1109/TCYB.2022.3155901
62. Li, Q., Tan, H., Wu, Y., Ye, L., Ding, F.: Traffic flow prediction with missing data imputed by tensor completion methods. IEEE Access. **8**, 63188–63201 (2020)

# Chapter 4
# Regularization and Momentum Coefficient-Free Non-negative Latent Factor Analysis via PSO

## 4.1 Overview

A big-data-related application [1–11] commonly involves numerous nodes with the inherent non-negativity interaction relationships, i.e., user-item interactions in a recommender system [12–14]. Due to the exponential growth of involved nodes, it is impossible to obtain their whole interaction relationships (e.g., a user touches a tiny subset of items only). Hence, a high-dimensional and sparse (HiDS) matrix [15–19] can described such inherent non-negativity interaction relationships, which has only a few known entries (describing the known interactions) while the most others are unknown rather than zeroes (describing the unknown ones).

Despite its extreme sparseness, an HiDS matrix contains rich knowledge regarding desired patterns like user communities [20], item clusters [21], topological and temporal neighbors [22], and user/item latent features [23]. Based on the analysis in Chaps. 2 and 3, a latent factor (LFA) model enables highly efficient and accurate extraction of essential features and knowledge from such an HiDS matrix. However, it mostly fails to fulfill non-negativity constraints [23, 24]. Note that an HiDS matrix from real application [1–5, 12–14] is commonly defined non-negative. Therefore, a non-negative model is needed to efficiently and accurately extract desired knowledge from it for various data analysis tasks [25–29].

Great efforts have been made for addressing this issue. Li et al. [30–33] have done a series of advanced non-negative matrix factorization (NMF) models to extract non-negative latent factors from a target matrix. Yu et al. [34] predict the comprehensive drug-drug interactions via semi-NMF. Pan et al. [35] embed an NMF model to boost the performance of mobile tracking. However, these models must fill the unknown entries of an HiDS matrix, thereby yielding unnecessarily high computational and storage cost.

To implement non-negative latent factor analysis (NLFA) on HiDS matrices efficiently, Luo et al. propose a single latent factor-dependent, non-negative and multiplicative update (SLF-NMU) algorithm to build an NLFA model, which only

Y. Yuan, X. Luo, *Latent Factor Analysis for High-dimensional and Sparse Matrices*, SpringerBriefs in Computer Science, https://doi.org/10.1007/978-981-19-6703-0_4

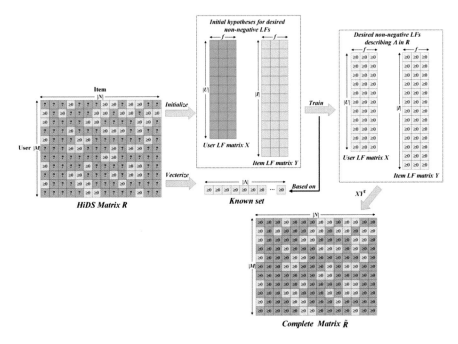

**Fig. 4.1** Flow diagram of an NLFA model for predictions of missing user preferences

depends the known part of an HiDS matrix. It greatly alleviates the computational and storage burden of performing NLFA on an HiDS matrix. Note that the computational cost of an NLFA model is linear with its known entry count only [36]. For better illustrating the work flow of an NLFA model, we give its prediction flow diagram in Fig. 4.1.

All the existing NLFA models' representative ability is limited due to their specialized learning objective, i.e., Euclidean distance. Note that Euclidean distance is only a special case of $\alpha$-$\beta$-divergence [33, 37], which have strong scalability and high robustness. Hence, if we model the objective function of an NLFA model with $\alpha$-$\beta$-divergence instead of Euclidean distance, can an NLFA model's representative ability be improved?

Aiming at addressing the issues, we propose a *Generalized* and *Adaptive LFA* (GALFA) model. Its main idea is to model the learning objective with $\alpha$-$\beta$-divergence, thereby significantly enhancing its representation learning ability to an HiDS matrix. The contributions of this chapter are threefold:

(a) A generalized learning objective for performing NLFA on an HiDS matrix, which is based on $\alpha$-$\beta$-divergence from a data density-oriented perspective;
(b) A single latent factor-dependent, non-negative, multiplicative and momentum-incorporated update (SLF-NM$^2$U) is adopted to accelerate the training process; and

(c) Regularization coefficient and momentum coefficient adaptation is implemented by following the principle of particle swarm optimization (PSO).

For validating the performance of the proposed GALFA model, we conduct experiments on six HiDS matrices to demonstrate its effectiveness.

The remainder of this chapter is organized as follows: Sect. 4.2 describes an SLF-NMU-based NLFA model. Section 4.3 presents our GALFA model in detail. Section 4.4 introduces the experimental results. Section 4.5 concludes this chapter.

## 4.2  An SLF-NMU-Based NLFA Model

As shown in Sect. 1.2.2, the objective function of an LFA model with Euclidean distance is formulated by:

$$\varepsilon(X, Y) = \sum_{r_{m,n} \in \Lambda} (r_{m,n} - \widehat{r}_{m,n})^2 + \lambda \|x_{m,\square}\|_2^2 + \lambda \|y_{n,\square}\|_2^2$$

$$= \sum_{r_{m,n} \in \Lambda} (r_{m,n} - \widehat{r}_{m,n})^2 + \lambda \sum_{d=1}^{f} x_{m,d}^2 + \lambda \sum_{d=1}^{f} y_{n,d}^2 \qquad (4.1)$$

where $\widehat{r}_{m,n} = \sum_{d=1}^{f} x_{m,d} y_{n,d}$, $\lambda$ denotes the regularization coefficient, $x_{m,}$ is the $m$-th row vector in $X$, $y_{n,}$ is the $n$-th row vector and $Y$, $r_{m,n}$, $x_{m,d}$ and $y_{n,d}$ denote the single elements of $R$, $X$ and $Y$, $\|\cdot\|_2$ calculates the $L_2$ norm of the enclosed vector.

However, most HiDS data are commonly inherent non-negative, which requires an LFA model fulfill the non-negativity constraints. Hence, (4.1) can be reformulated as:

$$\varepsilon(X, Y) = \sum_{r_{m,n} \in \Lambda} \left( \left( r_{m,n} - \sum_{d=1}^{f} x_{m,d} y_{n,d} \right)^2 + \lambda \sum_{d=1}^{f} \left( x_{m,d}^2 + y_{n,d}^2 \right) \right)$$

$$s.t. \ \forall m \in M, n \in N, d \in \{1, 2, \ldots, f\} : x_{m,d} \geq 0, y_{n,d} \geq 0 \qquad (4.2)$$

As indicated by prior research [12], SLF-NMU is an efficient algorithm to extract non-negative LFs from the known data of an HiDS matrix. It firstly applies the additive gradient descent (AGD) to each desired LF, resulting in the following update rule:

$$\underset{X,Y}{\text{argmin}}\varepsilon(X,Y)\overset{AGD}{\Longrightarrow}\begin{cases}x_{m,d}\leftarrow x_{m,d}-\eta_{m,d}\sum_{n\in\Lambda(m)}\left(\lambda x_{m,d}+y_{n,d}\tilde{r}_{m,n}-y_{n,d}r_{m,n}\right)\\[2ex]y_{n,d}\leftarrow y_{n,d}-\eta_{n,d}\sum_{m\in\Lambda(n)}\left(\lambda y_{n,d}+x_{m,d}\tilde{r}_{m,n}-x_{m,d}r_{m,n}\right)\end{cases}$$

$$(4.3)$$

where $\Lambda(m)$ and $\Lambda(n)$ denote the subsets of $\Lambda$ related to $m\in M$ and $n\in N$. $\eta_{m,d}$ and $\eta_{n,d}$ denotes the learning rates corresponding to $x_{m,d}$ and $y_{n,d}$, respectively.

For keeping resultant LFs non-negative, SLF-NMU manipulates $\eta_{m,d}$ and $\eta_{n,d}$ to cancel the negative terms in (4.3). Therefore, the learning rules for $x_{m,d}$ and $y_{n,d}$ is:

$$\underset{X,Y}{\text{argmin}}\varepsilon(X,Y)\overset{SLF-NMU}{\Longrightarrow}\begin{cases}x_{m,d}\leftarrow x_{m,d}\dfrac{\sum_{n\in\Lambda(m)}y_{n,d}r_{m,n}}{\sum_{n\in\Lambda(m)}y_{n,d}\tilde{r}_{m,n}+\lambda|\Lambda(m)|x_{m,d}}\\[4ex]y_{n,d}\leftarrow y_{n,d}\dfrac{\sum_{m\in\Lambda(n)}x_{m,d}r_{m,n}}{\sum_{m\in\Lambda(n)}x_{m,d}\tilde{r}_{m,n}+\lambda|\Lambda(n)|y_{n,d}}\end{cases}$$

$$(4.4)$$

## 4.3  The Proposed GALFA Model

### 4.3.1  A Generalized Learning Objective

Note that Euclidean distance is a special case of $\alpha$-$\beta$-divergence [38]. Therefore, we can generalize (4.1) with a common form of $\alpha$-$\beta$-divergence, thereby achieving the following problem formulation:

$$\varepsilon=\sum_{r_{m,n}\in\Lambda}\left(-\frac{1}{\alpha\beta}\left(r_{m,n}^{\alpha}\tilde{r}_{m,n}^{\beta}-\frac{\alpha}{\alpha+\beta}r_{m,n}^{\alpha+\beta}-\frac{\beta}{\alpha+\beta}r_{m,n}^{\alpha+\beta}\right)+\lambda\sum_{d=1}^{f}\left(x_{m,d}^{2}+y_{n,d}^{2}\right)\right)$$

$$s.t.\alpha>0,\beta>0,\forall m\in M,n\in N,d\in\{1,2,\ldots,f\}:x_{m,d}\geq0,y_{n,d}\geq0 \qquad (4.5)$$

With it, we achieve an $\alpha$-$\beta$-divergence learning objective for an NLFA model. Note that by substituting $\alpha=\beta=1$ into (4.4), we obtain a Euclidean distance-based learning objective (4.2).

## 4.3.2 A Generalized SLF-NMU-Based Learning Rule

For the generalized objective function (4.5), we design a generalized SLF-NMU-based learning rule for it. Note that the learning rules for LF matrices $X$ and $Y$ based on an SLF-NMU algorithm [15] can be achieved by analyzing its Karush-Kuhn-Tucker (KKT) conditions [39]. We define $H^{|M| \times f}$ and $K^{|N| \times f}$ be Lagrangian multipliers corresponding to the constraints $X \geq 0$ and $Y \geq 0$, respectively. Then we achieve the Lagrangian function corresponding to (4.5) as:

$$L = \varepsilon + tr(HX^T) + tr(K^T Y)$$

$$= \varepsilon + \sum_m \sum_d h_{m,d} x_{m,d} + \sum_n \sum_d \kappa_{n,d} y_{n,d} \tag{4.6}$$

Considering the partial derivatives of $L$ with respect to $x_{m,d}$ and $y_{n,d}$, we have the following deduction:

$$\begin{cases} \dfrac{\partial L}{\partial x_{m,d}} = \dfrac{\partial L}{\partial x_{m,d}} + h_{m,d} = \sum_{n \in N_m} \left( -\dfrac{1}{\alpha} r_{m,n}^\alpha \tilde{r}_{m,n}^{\beta-1} y_{n,d} + \dfrac{1}{\alpha} \tilde{r}_{m,n}^{\alpha+\beta-1} y_{n,d} + 2\lambda x_{m,d} \right) + h_{m,d} \\ \dfrac{\partial L}{\partial y_{n,d}} = \dfrac{\partial L}{\partial y_{n,d}} + \kappa_{n,d} = \sum_{m \in M_n} \left( -\dfrac{1}{\alpha} r_{m,n}^\alpha \tilde{r}_{m,n}^{\beta-1} x_{m,d} + \dfrac{1}{\alpha} \tilde{r}_{m,n}^{\alpha+\beta-1} x_{m,d} + 2\lambda y_{n,d} \right) + \kappa_{n,d} \end{cases}$$

$$\tag{4.7}$$

We have the following findings based on (4.6) and (4.7):

(a) The partial derivatives in (4.6) should be set at zero $L$; and
(b) The KKT condition of (4.7) is $h_{m,d} x_{m,d} = 0$ and $\kappa_{n,d} y_{n,d} = 0$.

By substituting the above findings into (4.6) and (4.7), we achieve that:

$$x_{m,d} \sum_{n \in N_m} \left( \dfrac{1}{\alpha} \tilde{r}_{m,n}^{\alpha+\beta-1} y_{n,d} + 2\lambda x_{m,d} \right) = x_{m,d} \sum_{n \in N_m} \dfrac{1}{\alpha} r_{m,n}^\alpha \tilde{r}_{m,n}^{\beta-1} y_{n,d}$$

$$y_{n,d} \sum_{m \in M_n} \left( \dfrac{1}{\alpha} \tilde{r}_{m,n}^{\alpha+\beta-1} x_{m,d} + 2\lambda y_{n,d} \right) = y_{n,d} \sum_{m \in M_n} \dfrac{1}{\alpha} r_{m,n}^\alpha \tilde{r}_{m,n}^{\beta-1} x_{m,d} \tag{4.8}$$

From (4.8), the update rules of $x_{m,d}$ and $y_{n,d}$ is given as:

$$\underset{X,Y}{\text{argmin}}\varepsilon(X,Y)\xrightarrow{SLF-NMU}\begin{cases} x_{m,d} \leftarrow x_{m,d} \dfrac{\displaystyle\sum_{n\in N_m} r_{m,n}^{\alpha}\tilde{r}_{m,n}^{\beta-1}y_{n,d}}{\displaystyle\sum_{n\in N_m}\tilde{r}_{m,n}^{\alpha+\beta-1}y_{n,d}+\alpha\lambda|N_m|x_{m,d}} \\[3em] y_{n,d} \leftarrow y_{n,d} \dfrac{\displaystyle\sum_{m\in M_n} r_{m,n}^{\alpha}\tilde{r}_{m,n}^{\beta-1}x_{m,d}}{\displaystyle\sum_{m\in M_n}\tilde{r}_{m,n}^{\alpha+\beta-1}x_{m,d}+\alpha\lambda|M_n|y_{n,d}} \end{cases} \tag{4.9}$$

Note that by setting $\alpha = \beta = 1$ into (4.9), we get the SLF-NMU-based learning rules for a Euclidean distance-based NLFA model.

### 4.3.3  Generalized-Momentum Incorporation

Although an SLF-NMU algorithm is specifically designed for performing NLFA on an HiDS matrix efficiently, it leads to slow convergence [2, 40]. Hence, we adopt a generalized momentum method to accelerate the training process. Given decision parameter $\theta$ and objective function $\varepsilon(\theta)$, a generalized momentum method [19, 38] updates $\theta$ at the $t$-th iteration as follows,

$$\begin{cases} a_0 = 0 \\ a_t = \kappa a_{t-1} - \left(\theta_t' - \theta_{t-1}\right) \\ \theta_t = \theta_{t-1} - a_t \end{cases} \tag{4.10}$$

where $\theta'_t$ denotes the predicted state of $\theta$ relying on the adopted algorithm after the $t$-th iteration, and $a_t$ denotes the state of $a_0$ after the $t$-th iteration.

Let $X_{t-1}$ and $Y_{t-1}$ be the states of $X$ and $Y$ after the $(t-1)$-th iteration, and $X'_t$ and $Y'_t$ be the expected states obtained by (4.9), respectively. Thus, $X'_t$ and $Y'_t$ can be obtained:

$$\left(X_t', Y_t'\right) = \underset{X,Y}{\overset{SLF-NMU}{\text{argmin}}}\ \varepsilon(X_{t-1}, Y_{t-1}) \tag{4.11}$$

Then we can achieve the update increment:

$$\Delta_t = \theta_t' - \theta_{t-1} = \begin{bmatrix} X_t' \\ Y_t' \end{bmatrix} - \begin{bmatrix} X_{t-1} \\ Y_{t-1} \end{bmatrix} \tag{4.12}$$

By substituting (4.12) into (4.10), he update velocity after the first training iteration is:

$$a_1 = \kappa a_0 - \Delta_1 = -\begin{bmatrix} X'_1 \\ Y'_1 \end{bmatrix} + \begin{bmatrix} X_0 \\ Y_0 \end{bmatrix} \tag{4.13}$$

where $X_0$ and $Y_0$ denote the initial state of $X$ and $Y$, i.e., randomly generated non-negative matrices as discussed in [14, 19, 22]. Then, $X_1$ and $Y_1$ is achieved:

$$\begin{bmatrix} X_1 \\ Y_1 \end{bmatrix} = \begin{bmatrix} X_0 \\ Y_0 \end{bmatrix} - a_1 = \begin{bmatrix} X_0 \\ Y_0 \end{bmatrix} + \begin{bmatrix} X'_1 \\ Y'_1 \end{bmatrix} - \begin{bmatrix} X_0 \\ Y_0 \end{bmatrix} = \begin{bmatrix} X'_1 \\ Y'_1 \end{bmatrix} \tag{4.14}$$

Analogously, we have:

$$a_1 = -\begin{bmatrix} X'_1 \\ Y'_1 \end{bmatrix} + \begin{bmatrix} X_0 \\ Y_0 \end{bmatrix} = -\begin{bmatrix} X_1 \\ Y_1 \end{bmatrix} + \begin{bmatrix} X_0 \\ Y_0 \end{bmatrix} \Rightarrow \begin{bmatrix} X_1 \\ Y_1 \end{bmatrix} = \begin{bmatrix} X_0 \\ Y_0 \end{bmatrix} - a_1 \tag{4.15}$$

where the update increment of the second iteration $\Delta_2$ is given as:

$$(X'_2, Y'_2) = \overset{SLF-NMU}{\underset{X,Y}{argmin}} \ \varepsilon(X_1, Y_1) \Rightarrow \Delta_2 = \begin{bmatrix} X'_2 \\ Y'_2 \end{bmatrix} - \begin{bmatrix} X_1 \\ Y_1 \end{bmatrix} \tag{4.16}$$

where $X'_2$ and $Y'_2$ denote $X$ and $Y$'s expected state after the second training iteration. Analogously, the update velocity $a_2$ is achieved:

$$a_2 = \kappa a_1 - \Delta_2 = -\kappa \left( \begin{bmatrix} X_1 \\ Y_1 \end{bmatrix} - \begin{bmatrix} X_0 \\ Y_0 \end{bmatrix} \right) - \left( \begin{bmatrix} X'_2 \\ Y'_2 \end{bmatrix} - \begin{bmatrix} X_1 \\ Y_1 \end{bmatrix} \right)$$
$$\Rightarrow \begin{bmatrix} X_2 \\ Y_2 \end{bmatrix} = \begin{bmatrix} X_1 \\ Y_1 \end{bmatrix} - a_2 = \begin{bmatrix} X'_2 \\ Y'_2 \end{bmatrix} + \kappa \left( \begin{bmatrix} X_1 \\ Y_1 \end{bmatrix} - \begin{bmatrix} X_0 \\ Y_0 \end{bmatrix} \right) \tag{4.17}$$

Note that situations in the third to $t$-th iterations are the same as that in the second iteration. Hence, we achieve the accelerated learning scheme as:

$$\begin{cases} t=1: \begin{bmatrix} X_1 \\ Y_1 \end{bmatrix} = \begin{bmatrix} X'_1 \\ Y'_1 \end{bmatrix}, \\ t=2: \begin{bmatrix} X_t \\ Y_t \end{bmatrix} = \begin{bmatrix} X'_t \\ Y'_t \end{bmatrix} + \kappa \left( \begin{bmatrix} X_{t-1} \\ Y_{t-1} \end{bmatrix} - \begin{bmatrix} X_{t-2} \\ Y_{t-2} \end{bmatrix} \right) \end{cases} \tag{4.18}$$

Finally, by substituting the learning rule (4.9) into (4.18), we get:

$$\underset{X,Y}{argmin} \varepsilon(X, Y) \overset{SLF-NMU \ with \ Generalized \ Momentum}{\Rightarrow}$$

$$
\begin{cases}
t=1: \begin{cases}
x_{m,d}_1 \leftarrow x_{m,d}_0 \dfrac{\sum\limits_{n\in N_m} r^\alpha_{m,n}{}_0 \tilde{r}^{\beta-1}_{m,n}{}_0 y_{n,d}{}_0}{\sum\limits_{n\in N_m} \tilde{r}^{\alpha+\beta-1}_{m,n}{}_0 y_{n,d}{}_0 + \alpha\lambda|N_m|x_{m,d}{}_0} \\[2em]
y_{n,d}_1 \leftarrow y_{n,d}_0 \dfrac{\sum\limits_{n\in N_m} r^\alpha_{m,n}{}_0 \tilde{r}^{\beta-1}_{m,n}{}_0 x_{m,d}{}_0}{\sum\limits_{m\in M_n} \tilde{r}^{\alpha+\beta-1}_{m,n}{}_0 x_{m,d}{}_0 + \alpha\lambda|M_n|y_{n,d}{}_0}
\end{cases} \\[6em]
t\geq 2: \begin{cases}
x_{m,d}_t \leftarrow x_{m,d}_{t-1} \dfrac{\sum\limits_{n\in N_m} r^\alpha_{m,n}{}_{t-1}\tilde{r}^{\beta-1}_{m,n}{}_{t-1} y_{n,d}{}_{t-1}}{\sum\limits_{n\in N_m} \tilde{r}^{\alpha+\beta-1}_{m,n}{}_{t-1} y_{n,d}{}_{t-1} + \alpha\lambda|N_m|x_{m,d}{}_{t-1}} + max\left\{0,\ \kappa\left(x_{m,d}_{t-1}-x_{m,d}_{t-2}\right)\right\} \\[2.5em]
y_{n,d}_1 \leftarrow y_{n,d}_{t-1} \dfrac{\sum\limits_{m\in M_n} r^\alpha_{m,n}{}_{t-1}\tilde{r}^{\beta-1}_{m,n}{}_{t-1} x_{m,d}{}_{t-1}}{\sum\limits_{m\in M_n} \tilde{r}^{\alpha+\beta-1}_{m,n}{}_{t-1} x_{m,d}{}_{t-1} + \alpha\lambda|M_n|y_{n,d}{}_{t-1}} + max\left\{0,\ \kappa\left(y_{n,d}_{t-1}-y_{n,d}_{t-2}\right)\right\}
\end{cases}
\end{cases}
$$

$$(4.19)$$

Note that $\kappa\left(x_{m,d}_{t-1} - x_{m,d}_{t-2}\right)$ and $\kappa\left(y_{n,d}_{t-1} - y_{n,d}_{t-2}\right)$ are non-negative projection to prevent LFs in $X$ and $Y$ from becoming negative.

### 4.3.4 Regularization and Momentum Coefficient Adaptation via PSO

The regularization coefficient $\lambda$ and momentum coefficient $\kappa$ in (4.19) decide the proposed GALFA model. In this chapter, we implement self-adaptation of them for excellent practicability via PSO [41, 42].

Firstly, we build $q$ particles in a two-dimension space to make up a swarm, where the $j$-th particle's position and velocity are defined as:

$$
v_j = [v_{j\lambda}, v_{j\kappa}], s_j = [s_{j\lambda}, s_{j\kappa}] \tag{4.20}
$$

where $v_{j\kappa}$ and $v_{j\lambda}$ denote the velocity of the dimensional learning rate and regularization coefficient, $s_{j\kappa}$ and $s_{j\lambda}$ indicate their velocity, respectively. Thus, we consider the problem as a two-dimensional vector optimization problem. Then by substituting (4.20) into the evolution rule of PSO in (1.3), we have that:

$$\forall j \in \{1, \ldots, q\}$$

$$: \begin{cases} \begin{bmatrix} v_{j\lambda}(t) \\ v_{j\kappa}(t) \end{bmatrix} = w \begin{bmatrix} v_{j\lambda}(t-1) \\ v_{j\kappa}(t-1) \end{bmatrix} + c_1 r_1 \left( pb_j(t-1) - \begin{bmatrix} s_{j\lambda}(t-1) \\ s_{j\kappa}(t-1) \end{bmatrix} \right) \\ \qquad\qquad + c_2 r_2 \left( gb(t-1) - \begin{bmatrix} s_{j\lambda}(t-1) \\ s_{j\kappa}(t-1) \end{bmatrix} \right) \\ \begin{bmatrix} s_{j\lambda}(t) \\ s_{j\kappa}(t) \end{bmatrix} = \begin{bmatrix} s_{j\lambda}(t-1) \\ s_{j\kappa}(t-1) \end{bmatrix} + \begin{bmatrix} v_{j\lambda}(t) \\ v_{j\kappa}(t) \end{bmatrix} \end{cases} \qquad (4.21)$$

where $pb_j$ and $gb$ denotes the best particle location of particle and the best position of the whole swarm, $w$ is the non-negative inertia constant, $c_1$ and $c_2$ are cognitive and social coefficients, $r_1$ and $r_2$ are two uniform random numbers in the range of [0, 1], respectively. Note that in this chapter, only one hyper-parameter needs to be adaptation. Hence, the particles are searched in a one-dimensional space only.

For making the whole swarm better fit the known set $\Lambda$, we adopt the following fitness function for the $j$-th particle:

$$F(j) = \sqrt{\sum_{r_{m,n} \in \Omega} (r_{m,n} - r^{\wedge}{}_{m,n})^2 / |\Omega|} \qquad (4.22)$$

The position and velocity of each particle must be constrained in a certain range:

$$\begin{cases} s_j(t) = \left[ s_{j\lambda}(t), \ s_{j\kappa}(t) \right] = \begin{cases} s_{j\lambda}(t) = min \left( \hat{\lambda}, \ max \left( \check{\lambda}, \ s_{j\lambda}(t) \right) \right) \\ s_{j\kappa}(t) = min \left( \hat{\kappa}, \ max \left( \check{\kappa}, \ s_{j\kappa}(t) \right) \right) \end{cases} \\ v_j(t) = \left[ v_{j\lambda}(t), \ v_{j\kappa}(t) \right] = \begin{cases} v_{j\lambda}(t) = min \left( \hat{v}_\lambda, \ max \left( \check{v}_\lambda, \ v_{j\lambda}(t) \right) \right) \\ v_{j\kappa}(t) = min \left( \hat{v}_\kappa, \ max \left( \check{v}_\kappa, \ v_{j\kappa}(t) \right) \right) \end{cases} \end{cases} \qquad (4.23)$$

We commonly have $\hat{v}=0.2 \times (\hat{s} - \check{s})$ and $\check{v} = -\hat{v}$ Note that $\forall j \in \{1, \ldots, q\}$, $s_j$ is linked with the same group of LF matrices. In the $j$-th sub-iteration of the $t$-th evolving iteration, $X$ and $Y$ at are trained as:

$$\underset{X,Y}{\text{argmin}}\varepsilon(X, Y)\overset{GSN}{\Longrightarrow}$$

$$\begin{cases} t=1: \begin{cases} x_{(1)m,d} \leftarrow x_{(1)m,d} \dfrac{\sum\limits_{n\in N_m} r_{m,n}^{\alpha}\,\tilde{r}_{(1)m,n}^{\beta-1}\,y_{(1)n,d}}{\sum\limits_{n\in N_m} \tilde{r}_{(1)m,n}^{\alpha+\beta-1}\,y_{(1)n,d} + \alpha\lambda|N_m|x_{(1)m,d}} \\[4ex] y_{(1)n,d} \leftarrow y_{(1)n,d} \dfrac{\sum\limits_{m\in M_n} r_{(1)m,n}^{\alpha}\,\tilde{r}_{(1)m,n}^{\beta-1}\,x_{(1)m,d}}{\sum\limits_{m\in M_n} \tilde{r}_{(1)m,n}^{\alpha+\beta-1}\,x_{(1)m,d} + \alpha\lambda|M_n|y_{(1)n,d}} \end{cases} \\[12ex] t\geq 2: \begin{cases} x_{(j)m,d} \leftarrow x_{(j)m,d} \dfrac{\sum\limits_{n\in N_m} r_{m,n}^{\alpha}\,\tilde{r}_{(j)m,n}^{\beta-1}\,y_{n,d}}{\sum\limits_{n\in N_m} \tilde{r}_{(j)m,n}^{\alpha+\beta-1}\,y_{(j)n,d} + \alpha\lambda|N_m|x_{(j)m,d}} + max\left\{0,\ \kappa\left(x_{(j)m,d}-x_{(j)m,d}\right)\right\} \\[4ex] y_{(j)n,d} \leftarrow y_{(j)n,d} \dfrac{\sum\limits_{m\in M_n} r_{m,n}^{\alpha}\,\tilde{r}_{(j)m,n}^{\beta-1}\,x_{(j)m,d}}{\sum\limits_{m\in M_n} \tilde{r}_{(j)m,n}^{\alpha+\beta-1}\,x_{(j)m,d} + \alpha\lambda|M_n|y_{(j)n,d}} + max\left\{0,\ \kappa\left(y_{(j)n,d}-y_{(j)n,d}\right)\right\} \end{cases} \end{cases}$$

$$(4.24)$$

where the subscript ($j$) on $X$ and $Y$ denotes that their current states are linked with the $j$-th particle, i.e., $s_{j\kappa}$ and $s_{j\lambda}$.

## 4.3.5  Algorithm Design and Analysis

| Algorithm GALFA | |
|---|---|
| **Input:** $U, I, \Lambda, f, \alpha, \beta, q, w, c_1, c_2$ | |
| **Operation** | **Cost** |
| **initialize** $X^{|M|\times f}, X_A^{|M|\times f}, X_B^{|M|\times f}, Y^{|N|\times f}, Y_A^{|N|\times f}, Y_B^{|N|\times f}$ non-negatively | $\Theta((|M|+|N|)\times f)$ |
| **initialize** $X_C^{|M|\times f} = X_D^{|M|\times f} = X, Y_C^{|N|\times f} = Y_D^{|N|\times f} = Y,$ | $\Theta((|M|+|N|)\times f)$ |
| **initialize** $r_1, r_2, n = 0$, *Max-training-round* $= T$, $\tilde{b}^{1\times 2}, s^{1\times 2}, s^{1\times 2}, s^{1\times 2}, v^{1\times 2}, v^{1\times 2}$ | $\Theta(2)$ |
| **initialize** $S^{q\times 2}, V^{q\times 2}, b^{q\times 2}$ | $\Theta(q\times 2)$ |
| **initialize** $F^q$ | $\Theta(q)$ |
| **while not** converge **and** $t \leq T$ **do** | $\times t$ |
|   **for** $j = 1$ **to** $q$ | $\times q$ |
|     **reset** $X_A = 0, X_B = 0, Y_A = 0, Y_B = 0$ | $\Theta((|M|+|N|)\times f)$ |
|     $(X, Y) = $ **UPDATE**$(U, I, \Lambda, f, \alpha, \beta, X, Y, X_A, Y_A, X_B, Y_B, X_C, Y_C, X_D, Y_D, S)$ | $T_{\text{UPDATE}}$ |
|     **set** $X_D = X_C, X_C = X, Y_D = Y_C, Y_C = Y$ | $\Theta((|M|+|N|)\times f)$ |

(continued)

**Algorithm GALFA**

| | |
|---|---|
| $F_j = fitness(s_{j,.})$ | $\Theta(1)$ |
| **end for** | – |
| $(S, V, pb, gb) = \textbf{SELF-ADAPTATION}(q, gb, F, pb, S, V, s⍰, s⌣, v⍰, v⌣)$ | $T_{\text{SELF-ADAPTATION}}$ |
| $t = t + 1$ | $\Theta(1)$ |
| **end while** | – |
| **Output:** $X, Y$ | |


**Procedure UPDATE**

**Input:** $M, N, \Lambda, f, \alpha, \beta, X, Y, X_A, Y_A, X_B, Y_B, X_C, Y_C, X_D, Y_D, S$

| Operation | Cost |
|---|---|
| **for each** $r_{m,n} \in \Lambda$ | $\times|\Lambda|$ |
| $\tilde{r}_{m,n} = \sum_{d=1}^{f} x_{m,d} y_{n,d}$ | $\Theta(f)$ |
| **for** $d = 1$ **to** $f$ | $\times f$ |
| $xa_{m,d} = xa_{m,d} + r_{m,n}^{\alpha} \tilde{r}_{m,n}^{\beta-1} y_{n,d}, xb_{m,d} = xb_{m,d} + \tilde{r}_{m,n}^{\alpha+\beta-1} y_{n,d} + \alpha s_{j,1} x_{m,d}$ | $\Theta(1)$ |
| $ya_{n,d} = ya_{n,d} + r_{m,n}^{\alpha} \tilde{r}_{m,n}^{\beta-1} x_{m,d}, yb_{n,d} = yb_{n,f} + \tilde{r}_{m,n}^{\alpha+\beta-1} x_{m,d} + \alpha s_{j,1} y_{n,d}$ | $\Theta(1)$ |
| **end for** | – |
| **end for** | – |
| **for** $m \in M$ | $\times|M|$ |
| **for** $d = 1$ **to** $f$ | $\times f$ |
| $x_{m,d} = x_{m,d}(xa_{m,d}/xb_{m,d}) + \max\{0, s_{j,2}(xc_{m,d}\text{-}xd_{m,d})\}$ | $\Theta(1)$ |
| **end for** | – |
| **end for** | – |
| **for** $n \in N$ | $\times|N|$ |
| **for** $d = 1$ **to** $f$ | $\times f$ |
| $y_{n,d} = y_{n,d}(ya_{n,d}/yb_{n,d}) + \max\{0, s_{j,2}(yc_{n,d}\text{-}yd_{n,d})\}$ | $\Theta(1)$ |
| **end for** | – |
| **end for** | – |
| **Output:** update $X, Y$ | |


**Procedure SELF-ADAPTATION**

**Input:** $q, gb, F, pb, S, V, s⍰, s⌣, v⍰, v⌣$

| Operation | Cost |
|---|---|
| **for** $j = 1$ **to** $q$ | $\times q$ |
| **if** $F_j < fitness(pb_{j,.})$ | $\Theta(1)$ |
| $fitness(pb_{j,.}) = F_j$ | $\Theta(1)$ |
| **for** $k = 1$ **to** $2$ | $\Theta(2)$ |
| $pb_{j,k} = s_{j,k}$ | $\Theta(1)$ |
| **end for** | – |
| **end if** | – |
| **if** $F_j < fitness(gb)$ | $\Theta(1)$ |
| $fitness(gb) = F_j$ | $\Theta(1)$ |

(continued)

**Procedure SELF-ADAPTATION**

| | |
|---|---|
| **for** $k = 1$ **to** 2 | $\Theta(2)$ |
| $gb_k = s_{j,k}$ | $\Theta(1)$ |
| **end for** | – |
| **end if** | – |
| **end for** | – |
| **for** $j = 1$ **to** $q$ | $\times q$ |
| **for** $k = 1$ **to** 2 | $\times 2$ |
| $v_{j,k} = wv_{j,k} + c_1 r_1(pb_{j,k}\text{-}s_{j,k}) + c_1 r_1(gb_k\text{-}s_{j,k})$ | $\Theta(1)$ |
| **if** $v_{j,k} > v\hat{}_k$ | $\Theta(1)$ |
| $v_{j,k} = v\hat{}_k$ | $\Theta(1)$ |
| **else if** $v_{j,k} < v\check{}_k$ | $\Theta(1)$ |
| $v_{j,k} = v\check{}_k$ | $\Theta(1)$ |
| **end if** | – |
| $s_{j,k} = s_{j,k} + v_{j,k}$ | $\Theta(1)$ |
| **if** $s_{j,k} > s\hat{}_k$ | $\Theta(1)$ |
| $s_{j,k} = s\hat{}_k$ | $\Theta(1)$ |
| **else if** $s_{j,k} < s\check{}_k$ | $\Theta(1)$ |
| $s_{j,k} = s\check{}_k$ | $\Theta(1)$ |
| **end if** | – |
| **end for** | – |
| **end for** | – |
| **Output: update** $S, V, pb, gb$ | |

According to the above analyses, we develop Algorithm GALFA, which uses several auxiliary matrices.

(a) For $X$, $X_A$ and $X_B$ are adopted to cache the training increment on each instance $r_{m,n} \in \Lambda$, and $X_C$ and $X_D$ to cache the intermediate results of the $(t-1)$-th and $(t-2)$-th iterations. Similar design is also applied to $Y$;

(b) To implement self-adaptation of $\lambda$ and $\kappa$, $S$, $V$, $P_B$ and $F$ are adopted to cache the updated position, updated velocity, best position, fitness function value of particles, respectively.

Note that Algorithm GALFA relies on procedure UPDATE and SELF-ADAPTATION. According to them, we see that their costs are $T_{\text{UPDATE}} \approx \Theta((|M| + |N|) \times f + |\Lambda| \times f)$ and $T_{\text{SELF-ADAPTATION}} \approx q$, respectively. Hence, Algorithm GALFA's cost is:

$$
\begin{aligned}
T &\approx t \cdot \Theta(q \times ((|M| + |N|) \times f + |\Lambda| \times f) + q) \\
&\approx \Theta(t \times q \times |\Lambda| \times f).
\end{aligned}
\tag{4.25}
$$

Note that in (4.25) we adopt the condition of $|\Lambda| \ll \max\{|M|, |N|\}$ to reasonably omit the lower-order-terms. Thus, Algorithm GALFA's computational complexity is linear with $|\Lambda|$ since $t$, $q$ and $f$ are both positive constants.

Meanwhile, we see that Algorithm GALFA uses auxiliary matrices, i.e., $X$, $Y$, $X_A$, $Y_A$, $X_B$, $Y_B$, $X_C$, $Y_C$, $X_D$, $Y_D$, $S$, $V$, $P_B$ and $F$. Therefore, we have its storage cost as:

$$
\begin{aligned}
S & = 5 \times (|M| + |N|) \times f + 7 \times q + |\Lambda| \\
  & \approx 5 \times (|M| + |N|) \times f + |\Lambda|
\end{aligned}
\tag{4.26}
$$

Based on (4.25) and (4.26), Algorithm GALFA has high efficiency in both storage and computation.

## 4.4  Experimental Results and Analysis

### 4.4.1  General Settings

**Evaluation Protocol**  We adopt estimation for missing data of an HiDS matrix as the evaluation protocol for the experiments due to its popularity. It is commonly measured by root mean squared error (RMSE) [3, 4, 14, 19, 43, 44]:

$$
RMSE = \sqrt{\left( \sum_{r_{m,n} \in \Phi} (r_{m,n} - \tilde{r}_{m,n})^2 \right) / |\Phi|},
\tag{4.27}
$$

where $|\cdot|$ calculates the cardinality of an enclosed set.

**Datasets**  Six HiDS matrices adopted in our experiments, which is recorded in Table 4.1.

Note that each dataset is randomly split into ten disjoint subsets for implementing 80%–20% train-test settings. The above process is repeated for ten times to achieve the final averaged results. Considering each model in a single run, the training process breaks if (a) the iteration count reaches 1000; or (b) the RMSE difference in two consecutive iterations is within $5 \times 10^{-6}$.

**Model Settings**  To obtain a fair and objective comparison, we adopt the following measures settings:

**Table 4.1**  Datasets details

| No. | Name | Row | Column | Known Entries | Density (%) |
|-----|------|-----|--------|---------------|-------------|
| D1 | MovieLens 20M | 26,744 | 138,493 | 20,000,263 | 0.54 |
| D2 | Douban | 58,541 | 129,490 | 16,830,839 | 0.22 |
| D3 | EachMovie | 72,916 | 1628 | 2,811,718 | 2.37 |
| D4 | Flixter | 48,794 | 147,612 | 8,196,077 | 0.11 |
| D5 | Dating | 135,359 | 168,791 | 17,359,346 | 0.076 |
| D6 | Epinion | 120,492 | 775,760 | 13,668,320 | 0.015 |

(a) LF matrices $X$ and $Y$ are initialized non-negative randomly generated for eliminating the performance bias; and

(b) LF dimension $f = 20$ to balance the representative learning ability and computational efficiency of each model.

(c) For PSO, the dimension space $D = 2$, the search space of the regularization coefficient is [0, 1.4], and the search space of the momentum coefficient is [0.02, 0.12]. $r_1$ and $r_2$ are two uniformly distributed random numbers in a range of [0, 1], $c_1 = c_2 = 2$, $S = 10$, $w = 0.729$ according to [45].

### 4.4.2   Parameter Sensitivity

1. *Effects of $\kappa$ and $\lambda$'s adaptation*

   In this part, we aim at validating the performance of $\lambda$ and $\kappa$'s self-adaptation in GALFA. We manually tune them to achieve the optimal outputs. This process is depicted in Figs. 4.2 and 4.3. From them, we have the following findings:

   (a) GALFA's performance is affected by the regularization coefficient and momentum coefficient. As depicted in Fig. 4.2, on D1, the RMSE with $\lambda = 0.02, 0.04, 0.06, 0.08, 0.10$ and $0.12$ is $0.7894, 0.7826, 0.7869, 0.7965, 0.8091$ and $0.8213$ when we set $\kappa = 1$. We can find that the RMSE is $0.7826$

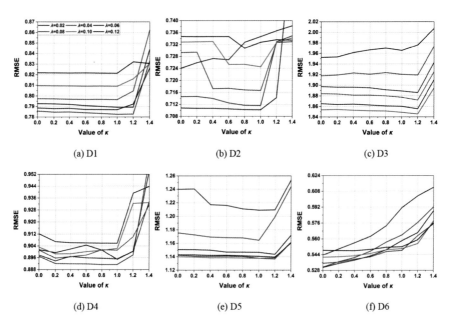

**Fig. 4.2** Performance of GALFA as $\lambda$ and $\kappa$ vary without self-adaptation. All panels share the legend in panel (**a**)

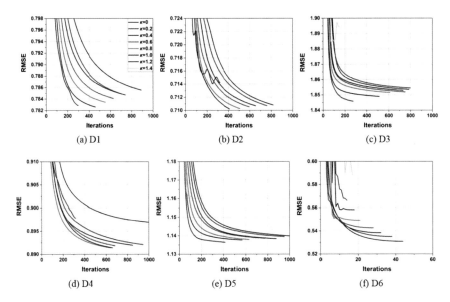

**Fig. 4.3**  Training process of GALFA with optimal $\lambda$ and $\kappa$ changing. All panels share the legend in panel (**a**)

with $\lambda = 0.04$ and 0.8213 with $\lambda = 0.12$. The accuracy gain reaches 4.71%. Similar outcomes are also found on other testing cases.

(b) GALFA's convergence rate is significantly improved by a generalized momentum method. Note that large $\kappa$ makes a learning algorithm overshoot a local optimum. For instance, on D3, GALFA achieves the RMSE at 1.8943 with $\kappa = 1.4$, 2.56% higher than 1.8458 with $\kappa = 1.2$. In contrast, a small $\kappa$ has no obvious momentum effects. For instance, on D3, GALFA consumes 798 iterations to achieve RMSE at 1.8544 with $\kappa = 0.2$. In contrast, with $\kappa = 1.2$, GALFA only consumes 275 iterations to achieve the RMSE at 1.8458. The number of iterations decreases at 65.54%. Similar results are found on other testing cases. However, the situation is different on D6. GALFA' prediction accuracy keeps decreasing with momentum effects.

Moreover, Table 4.2 records the model performance with self-adapted and manually-tuned hyper-parameters. The training process of GALFA with adaptive $\kappa$ and $\lambda$ is depicted in Fig. 4.4. As shown in Table 4.2, optimal $\lambda$ and $\kappa$ are data dependence.

(a) $\kappa$ and $\lambda$ adaptation affects no prediction accuracy of GALFA. As shown in Table 4.2, on D2, D3 and D5, GALFA with self-adaptation of $\kappa$ and $\lambda$ achieves lower RMSE than one with manually-tuned one. For instance, on D4, GALFA's RMSE is 0.8914 with $\kappa$ and $\lambda$ adaptation, and 0.8919 with manually-tuned. Considering other datasets, $\kappa$ and $\lambda$ adaptation results in slight prediction

**Table 4.2** Model performance with self-adapted and manually-tuned $\lambda$ and $\kappa$

| Dataset | Best $\lambda$ and $\kappa$ | RMSE | Iterations | Per | Total |
|---------|------------------|--------|------------|--------|--------|
| D1 | 0.04, 1.0 | **0.7826** | 461 | **6.485** | 2989.6 |
|  | Self-adaptation | 0.7839 | **40** | 65.23 | **2609.2** |
| D2 | 0.06, 1.0 | 0.7103 | 411 | **5.779** | 2375.2 |
|  | Self-adaptation | **0.7100** | **32** | 57.93 | **1853.7** |
| D3 | 0.10, 1.2 | 1.8458 | 275 | **7.861** | 2161.8 |
|  | Self-adaptation | **1.8432** | **16** | 79.22 | **1267.5** |
| D4 | 0.04, 0.8 | **0.8914** | 644 | **2.748** | 1769.7 |
|  | Self-adaptation | 0.8918 | **27** | 27.98 | **755.5** |
| D5 | 0.10, 1.2 | 1.1364 | 414 | **8.099** | 3352.9 |
|  | Self-adaptation | **1.1356** | **34** | 80.12 | **2724.1** |
| D6 | 0.06, 0.0 | **0.5309** | 44 | **5.818** | 255.9 |
|  | Self-adaptation | 0.5316 | **3** | 58.44 | **175.3** |

"Per" denotes Time cost per iteration (seconds), and "Total" denotes Total time cost (seconds).

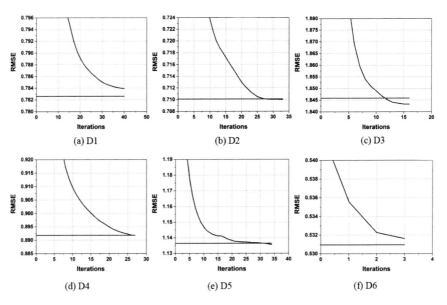

**Fig. 4.4** Training process of GALFA with self-adapted $\kappa$ and $\lambda$. Redline denotes the RMSE of GALFA with manually-tuned

accuracy loss. For instance, on D6, GALFA's RMSE is 0.5309 with $\kappa$ and $\lambda$ adaptation, and 0.5316 with manually-tuned. The accuracy loss is only 0.13%.

(b) $\kappa$ and $\lambda$ adaptation greatly reduces the total training time cost. From Table 4.2 we see that owing to $\kappa$ and $\lambda$ adaptation, GALFA's iterations decreases drastically. Hence, compared with the model with manually tuned $\kappa$ and $\lambda$, its total time cost decreases significantly. For instance, as recorded in Table 4.2, with/without $\kappa$

**Table 4.3** GALFA's RMSE with the best $\alpha$ and $\beta$ V.S. $\alpha = \beta = 1$ (i.e., Euclidean distance)

| Dataset | Best $\alpha$ and $\beta$ | RMSE | RMSE as $\alpha = \beta = 1$ | Gap (%) |
|---------|---------------------------|------|------------------------------|---------|
| D1 | $\alpha = 1.2, \beta = 0.8$ | **0.7811** | 0.7839 | 0.36 |
| D2 | $\alpha = 1.0, \beta = 0.8$ | **0.7058** | 0.7100 | 0.59 |
| D3 | $\alpha = 1.0, \beta = 0.4$ | **1.8187** | 1.8432 | 1.33 |
| D4 | $\alpha = 1.0, \beta = 0.8$ | **0.8782** | 0.8918 | 1.53 |
| D5 | $\alpha = 1.2, \beta = 1.0$ | **1.1277** | 1.1356 | 0.70 |
| D6 | $\alpha = 1.0, \beta = 0.4$ | **0.5212** | 0.5316 | 1.96 |

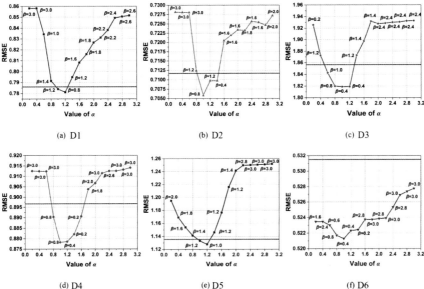

**Fig. 4.5** RMSE of GALFA as $\alpha$ and $\beta$ vary. Redline denotes the RMSE as $\alpha = \beta = 1.0$, i.e., Euclidean distance

and $\lambda$ adaptation, on D4, GALFA consumes 755.5 and 1769.7 s, respectively. Thus, its time cost reduces 57.31%. The similar conclusions can be made on the other testing cases. Moreover, Table 4.2 only records its time cost with the pre-tuned $\kappa$ and $\lambda$, but does not record the time cost to tune them. Since such tuning requires a twofold grid search, the cost is much more than the training cost. From this point of view, GALFA's time efficiency is greatly enhanced with $\kappa$ and $\lambda$ adaptation.

2. *Effects of $\alpha$ and $\beta$*

   In this part, we validate $\alpha$ and $\beta$ effects. Note that as analyzed before, the $\alpha$-$\beta$-divergence degenerates to the Euclidean distance when $\alpha = \beta = 1$. Table 4.3 records GALFA's RMSE with the best $\alpha$ and $\beta$ versus $\alpha = \beta = 1$. Figure 4.5 shows the RMSE as $\alpha$ and $\beta$ change. From them, we have the following important findings:

(a) GALFA's prediction accuracy is closely connected with $\alpha$ and $\beta$. Note that the commonly adopted Euclidean distance is not the best choice on testing cases. For instance, as recorded in Table 4.3, on D1, the lowest RMSE is 0.7811 with the best $\alpha = 1.2$ and $\beta = 0.8$. In contrast, the case of Euclidean distance achieves the lowest RMSE at 0.7839. The accuracy gain is 0.36% with the optimized $\alpha$-$\beta$-divergence. Similar outcomes also exist on the other testing cases.

(b) The adaptaion of $\alpha$ and $\beta$ is highly desired. As shown in Table 4.3, we see that the optimal values of $\alpha$ and $\beta$ are data-dependent. However, the adaptation strategy of $\kappa$ and $\lambda$ is unsuited for $\alpha$ and $\beta$. The main reason is that $\kappa$ and $\lambda$ are only controllable parameters in a single model, while $\alpha$ and $\beta$ are modeling parameters affecting the fundamental distance metric of our learning objective.

### 4.4.3  Comparison with State-of-the-Art Models

In this part of the experiments, we compare the proposed GALAF model with several state-of-the-art models on estimation accuracy and computational efficiency for missing data of an HiDS matrix. The details of all the compared models are summarizes in Table 4.4 as followings:

Figures 4.6 and 4.7 depict the RMSE and the corresponding total time cost, respectively. From them, we have the following important findings:

(a) GALFA accurately recovers missing data of an HiDS matrix. For instancae, On D2 and D4–6, M1 is able to achieve lower RMSE than its peers do. For instance, as depicted in Fig. 4.6, on D4, M1's RMSE is 0.8782, which is 2.08% lower than 0.8969 by M2, 1.92% lower than 0.8954 by M3, 1.75% lower than 0.8938 by M4, 0.34% lower than 0.8812 by M5, 1.56% lower than 0.8921 by M6, 0.11% lower than 0.8792 by M7, and 0.97% lower than 0.8865 by M8. Note that Note that M3–5 and M7 are state-of-the-art deep-learning-incorporated models with

**Table 4.4**  Details of compared models

| No. | Model | Description |
|-----|-------|-------------|
| M1 | GALFA | The proposed model of this study |
| M2 | NLFA | An SLF-NMU-based NLFA model [40] |
| M3 | NeuMF | A deep neural network-based LF model [37] |
| M4 | I-AutoRec | An autoencoder paradigm-based LF model [46] |
| M5 | $\beta$-NLFA | An NLFA model based on a $\beta$-divergence function [2] |
| M6 | DCCR | An autoencoder and a multilayered perceptron-based LFA model [47] |
| M7 | NRT | A recurrent neural network-based LFA model [13] |
| M8 | FBNLFA | A linear biases-incorporated fast NLFA model |

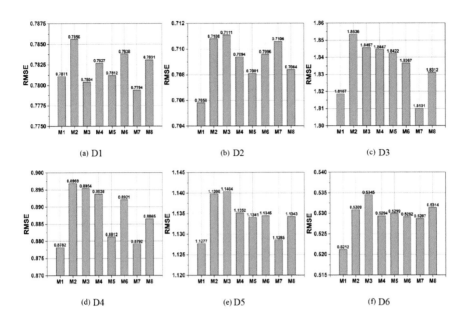

**Fig. 4.6**  RMSE of M1–8

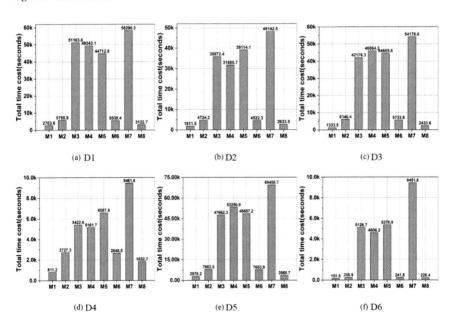

**Fig. 4.7**  Total Time cost of M1–8

**Table 4.5** Average ranks of all compared models

| Rank | M1 | M2 | M3 | M4 | M5 | M6 | M7 | M8 |
|------|------|------|------|------|------|------|------|------|
| Accuracy | 1.50 | 7.16 | 6.50 | 5.00 | 3.83 | 5.33 | 2.33 | 4.33 |
| Efficiency | 1.00 | 4.00 | 5.83 | 5.83 | 6.33 | 3.00 | 8.00 | 2.00 |

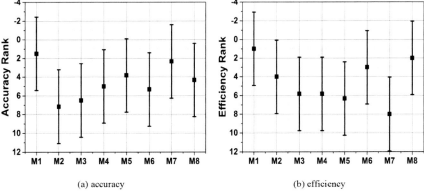

(a) accuracy                                    (b) efficiency

**Fig. 4.8** Results of Nemenyi analysis

high representative learning ability. Thus, M1's ability to estimate an HiDS matrix's missing data is impressive.

(b) GALFA's computational cost is the lowest among its peers. For instance, as depicted in Fig. 4.7, on D1, M1 consumes 2703.6 s to converge, which is 46.97% of 5755.9 s by M2, 5.28% of 51163.6 s by M3, 5.48% of 49343.1 s by M4, 6.04% of 44712.8 s by M5, 48.82% of 5538.4 s by M6, 4.64% of 58298.3 s by M7, and 86.58% of 3122.7 s by M8. The similar outcomes are also seen on other testing cases. The main reason is that for the deep neural network-based models, i.e., M3–5 and M7, they perform lots of matrix operations to implement backward propagation (BP)-based training. Therefore, their computational efficiency is low when addressing an HDI matrix even with GPU-based acceleration [8, 9]. For M2 and M6, they do not have the acceleration design as that of M1. Hence, they consume much more iterations to converge than M1 does, which greatly increases their time cost. Although M8 is accelerated by a generalized momentum method, M1 makes hyper-parameters self-adaptation to further reduce the time cost to train GALFA.

(c) Significance Analysis. Friedman test is adopted to validate the statistical significance of the above results. The Friedman statistical results are recorded in Table 4.5. From it, we clearly see that GALFA achieves the lowest Rank value, which means it outperforms other compared models in terms of computational efficiency and estimation accuracy. For further identifying two models are significantly different, a Nemenyi test is adopted. The results of Nemenyi analysis are depicted in Fig. 4.8. As drawn in Fig. 4.8a, we see that considering prediction accuracy, M1 outperforms M2–8. Especially, it significantly outperforms M2 and M3. Considering computational efficiency, M1 is always steadily consumes the least total time cost among all tested models. Meanwhile, as illustrated in Fig. 4.8b, the computational efficiency of M1 is significantly higher than M3–5 and M7.

### 4.4.4 Summary

According to the experimental results, we summarize that:

(a) With carefully tuned $\alpha$ and $\beta$, it achieves high prediction accuracy for missing data of an HiDS matrix.
(b) Its computational efficiency is significantly higher than state-of-the-art models owing to its generalized momentum-incorporated parameter learning rules and self-adaptation of controllable hyper-parameters.

## 4.5 Conclusions

This study proposes GALFA for performing NLFA on an HiDS matrix. Compared with the state-of-the-art models, it enjoys (a) high ability to represent an HiDS matrix owing to its $\alpha$-$\beta$-divergence-generalized learning objective, and (b) fast convergence relying on its generalized momentum-incorporated learning scheme.

The proposed GALFA has two types of hyper-parameters, i.e., controllable hyper-parameters and model hyper-parameters. We have incorporated the principle of PSO to make the controllable hyper-parameters self-adaptation. However, a standard PSO algorithm often suffers from premature convergence [39]. Therefore, incorporating improved PSO, e.g., aging mechanism PSO [48], genetic learning PSO [49], and quantum mechanism PSO [50], can help GALFA achieve further performance gain. On the other hand, model hyper-parameters decide the detailed form of GALFA's generalized learning objective. They presently require manual tuning. Since they affect GALFA from a model's perspective, their adaptation strategies are not compatible with that of controllable hyper-parameters. How to implement their self-adaptation remains open. We plan to address it as our future work.

# References

1. Resnick, P., Varian, H.R.: Recommender systems. Commun. ACM. **40**(3), 56–59 (1997)
2. Luo, X., Yuan, Y., Zhou, M.C., Liu, Z.G., Shang, M.S.: Non-negative latent factor model based on $\beta$-divergence for recommender systems. IEEE Trans. Syst. Man Cybern. Syst. **51**(8), 4612–4623 (2019). https://doi.org/10.1109/TSMC.2019.2931468
3. Sha, J., Du, Y.Y., Qi, L.: A user requirement oriented web service discovery approach based on logic and threshold petri net. IEEE/CAA J. Autom. Sin. **6**(6), 1528–1542 (2019)
4. Zahid, H., Mahmood, T., Morshed, A., Sellis, T.: Big data analytics in telecommunications: literature review and architecture recommendations. IEEE/CAA J. Autom. Sin. **7**(1), 18–38 (2020)
5. Chen, J., Lian, D., Zheng, K.: Improving one-class collaborative filtering via ranking-based implicit regularizer. In: Proc. of the 33rd AAAI Conf. on Artificial Intelligence, pp. 37–44 (2019)
6. Hu, L., Yan, S., Luo, X., Zhou, M.C.: An algorithm of inductively identifying clusters from attributed graphs. IEEE Trans. Big Data. https://doi.org/10.1109/TBDATA.2020.2964544
7. Qi, Y., Jin, L., Luo, X., Zhou, M.C.: Recurrent neural dynamics models for perturbed nonstationary quadratic programs: a control-theoretical perspective. IEEE Trans. Neural Netw. Learn. Syst. https://doi.org/10.1109/TNNLS.2020.3041364
8. Hu, L., Hu, P., Yuan, X., Luo, X., You, Z.: Incorporating the coevolving information of substrates in predicting HIV-1 protease cleavage sites. IEEE/ACM Trans. Comput. Biol. Bioinform. **17**(6), 2017–2028 (2020)
9. Wei, L., Jin, L., Luo, X.: Noise-suppressing neural dynamics for time-dependent constrained nonlinear optimization with applications. IEEE Trans. Syst. Man Cybern. Syst. https://doi.org/10.1109/TSMC.2021.3138550
10. Cheng, D., Huang, J., Zhang, S., Zhang, X., Luo, X.: A novel approximate spectral clustering algorithm with dense cores and density peaks. IEEE Trans. Syst. Man Cybern. Syst. https://doi.org/10.1109/TSMC.2021.3049490
11. Zhong, Y.R., Jin, L., Shang, M.S., Luo, X.: Momentum-incorporated symmetric non-negative latent factor models. IEEE Trans. Big Data. https://doi.org/10.1109/TBDATA.2020.3012656
12. Luo, X., Zhou, M.C., Wang, Z.D., Xia, Y.N., Zhu, Q.S.: An effective scheme for QoS estimation via alternating direction method-based matrix factorization. IEEE Trans. Serv. Comput. **12**(4), 503–518 (2019)
13. Li, P., Wang, Z., Ren, Z., Bing, L., Lam, W.: Neural rating regression with abstractive tips generation for recommendation. In: Proc. of the 40th Int. Conf. on Research and Development in Information Retrieval, pp. 345–354 (2017)
14. Zhang, S., Yao, L., Sun, A., Tay, Y.: Deep learning based recommender system: a survey and new perspectives. ACM Comput. Surv. **52**(1), 1–5 (2019)
15. Luo, X., Liu, Z.G., Li, S., Shang, M.S., Wang, Z.D.: A fast non-negative latent factor model based on generalized momentum method. IEEE Trans. Syst. Man Cybern. Syst. **51**(1), 610–620 (2019). https://doi.org/10.1109/TSMC.2018.2875452
16. Luo, X., Zhou, M.C., Li, S., Hu, L., Shang, M.S.: Non-negativity constrained missing data estimation for high-dimensional and sparse matrices from industrial applications. IEEE Trans. Cybern. **50**(5), 1844–1855 (2018). https://doi.org/10.1109/TCYB.2018.2894283
17. Luo, L., Xiong, Y., Liu, Y., Sun, X.: Adaptive gradient methods with dynamic bound of learning rate. In: Proc. of the Int. Conf. on learning representations (2019)
18. Wu, D., Luo, X.: Robust latent factor analysis for precise representation of high-dimensional and sparse data. IEEE/CAA J. Autom. Sin. **8**(4), 796–805 (2021)
19. Luo, X., Sun, J.P., Wang, Z.D., Li, S., Shang, M.S.: Symmetric and non-negative latent factor models for undirected, high dimensional and sparse networks in industrial applications. IEEE Trans. Ind. Inform. **13**(6), 3098–3107 (2017)
20. Yuan, Y., Luo, X., Shang, M.: Effects of preprocessing and training biases in latent factor models for recommender systems. Neurocomputing. **275**, 2019–2030 (2018)

21. Nathanson, T., Bitton, E., Goldberg, K.: Eigentaste 5.0: constant-time adaptability in a recommender system using item clustering. In: Proc. of the ACM Conf. on Recommender systems, pp. 149–152 (2007)

22. Rafailidis, D., Nanopoulos, A.: Modeling users preference dynamics and side information in recommender systems. IEEE Trans. Syst. Man Cybern. Syst. **46**(6), 782–792 (2016)

23. Luo, X., Zhou, M., Li, S., Shang, M.: An inherently nonnegative latent factor model for high-dimensional and sparse matrices from industrial applications. IEEE Trans. Ind. Inform. **14**(5), 2011–2022 (2018)

24. Liu, Z., Yuan, G., Luo, X.: Symmetry and nonnegativity-constrained matrix factorization for community detection. IEEE/CAA J. Autom. Sin. **9**(9), 1691–1693. https://doi.org/10.1109/JAS.2022.1005794

25. Chen, J., Wang, R., Wu, D., Luo, X.: A differential evolution-enhanced position-transitional approach to latent factor analysis. IEEE Trans. Emerg. Topics Comput. Intell. https://doi.org/10.1109/TETCI.2022.3186673

26. Wu, D., Zhang, P., He, Y., Luo, X.: A double-space and double-norm ensembled latent factor model for highly accurate web service QoS prediction. IEEE Trans. Serv. Comput. https://doi.org/10.1109/TSC.2022.3178543

27. Yuan, Y., Luo, X., Shang, M., Wang, Z.: A Kalman-filter-incorporated latent factor analysis model for temporally dynamic sparse data. IEEE Trans. Cybern. https://doi.org/10.1109/TCYB.2022.3185117

28. Qin, W., Wang, H., Zhang, F., Wang, J., Luo, X., Huang, T.: Low-rank high-order tensor completion with applications in visual data. IEEE Trans. Image Process. https://doi.org/10.1109/TIP.2022.3155949

29. Chen, X., Luo, X., Jin, L., Li, S., Liu, M.: Growing echo state network with an inverse-free weight update strategy. IEEE Trans. Cybern. https://doi.org/10.1109/TCYB.2022.3155901

30. Li, Z.C., Tang, J.H., Zhang, L.Y., Yang, J.: Weakly-supervised semantic guided hashing for social image retrieval. Int. J. Comput. Vis. (2020). https://doi.org/10.1007/s11263-020-01331-0

31. Li, Z.C., Tang, J.H., Mei, T.: Deep collaborative embedding for social image understanding. IEEE Trans. Pattern Anal. Mach. Intell. **41**(9), 2070–2083 (2019)

32. Li, Z.C., Tang, J.H., He, X.F.: Robust structured nonnegative matrix factorization for image representation. IEEE Trans. Neural Netw. Learn. Syst. **29**(5), 1947–1960 (2018)

33. Li, Z.C., Tang, J.H.: Weakly-supervised deep matrix factorization for social image understanding. IEEE Trans. Image Process. **26**(1), 276–288 (2017)

34. Yu, H., Mao, K.T., Shi, J.Y., Huang, H., Chen, Z., Dong, K., Yiu, S.M.: Predicting and understanding comprehensive drug-drug interactions via semi-nonnegative matrix factorization. BMC Syst. Biol. **12**(1), 101–110 (2018)

35. Pan, J.J., Pan, S.J., Jie, Y., Ni, L.M., Qiang, Y.: Tracking mobile users in wireless networks via semi-supervised co-localization. IEEE Trans. Pattern Anal. Mach. Intell. **34**(3), 587–600 (2012)

36. Luo, X., Wang, Z.D., Shang, M.S.: An instance-frequency-weighted regularization scheme for non-negative latent factor analysis on high dimensional and sparse data. IEEE Trans. Syst. Man Cybern. Syst. (2019). https://doi.org/10.1109/TSMC.2019.2930525

37. He, X., Liao, L., Zhang, H., et al.: Neural collaborative filtering. In: Proc. Of the 26th Int. Conf. on World Wide Web, pp. 173–182 (2017)

38. Cichocki, A., Cruces, S., Amari, S.: Generalized alpha-beta divergences and their application to robust nonnegative matrix factorization. Entropy. **13**(1), 134–170 (2011)

39. Gao, S., Zhou, M.Z., Wang, Y., Cheng, J., Yachi, H., Wang, J.: Dendritic neuron model with effective learning algorithms for classification, approximation and prediction. IEEE Trans. Neural Netw. Learn. Syst. **30**(2), 601–614 (2019)

40. Luo, X., Zhou, M., Xia, Y., Zhu, Q.: An efficient non-negative matrix-factorization-based approach to collaborative-filtering for recommender systems. IEEE Tran. Ind. Inform. **10**(2), 1273–1284 (2014)

41. Li, J., Zhang, J., Jiang, C., Zhou, M.: Composite particle swarm optimizer with historical memory for function optimization. IEEE Trans. Cybern. **45**(10), 2350–2363 (2015)

42. Dong, W., Zhou, M.: A supervised learning and control method to improve particle swarm optimization algorithms. IEEE Trans. Syst. Man Cybern. Syst. **47**(7), 1149–1159 (2017)
43. Wu, D., Luo, X., Wang, G.Y., Shang, M.S., Yuan, Y., Yan, H.Y.: A highly-accurate framework for self-labeled semi-supervised classification in industrial applications. IEEE Trans. Ind. Inform. **43**(3), 909–920 (2018)
44. Jin, L., Li, S., La, H.M., Luo, X.: Manipulability optimization of redundant manipulators using dynamic neural networks. IEEE Trans. Ind. Electron. **64**(6), 4710–4720 (2017)
45. Wang, J.J., Kumbasar, T.: Parameter optimization of interval Type-2 fuzzy neural networks based on PSO and BBBC methods. IEEE/CAA J. Autom. Sin. **6**(1), 247–257 (2019)
46. S. Sedhain, A. K. Menon, S. Sanner, L. Xie, Autorec: autoencoders meet collaborative filtering, In: Proc. of the 24th Int. Conf. on World Wide Web, pp. 111–112 (2015)
47. Wang, Q., Peng, B., Shi, X., Shang, T., Shang, M.: DCCR: deep collaborative conjunctive recommender for rating prediction. IEEE Access. **7**, 60186–60198 (2019)
48. Chen, W.N., Zhang, J., Lin, Y., Chen, N., Zhan, Z.H., Chung, H., Shi, Y.H.: Particle swarm optimization with an aging leader and challengers. IEEE Trans. Evol. Comput. **17**(2), 241–258 (2013)
49. Gong, Y.J., Li, J.J., Zhou, Y.C., Li, Y., Chung, H., Shi, Y.H., Zhang, J.: Genetic learning particle swarm optimization. IEEE Trans. Cybern. **46**(10), 2277–2290 (2016)
50. Wang, H., Qiao, C.M.: A nodes' evolution diversity inspired method to detect anomalies in dynamic social networks. IEEE Trans. Knowl. Data Eng. (2019). https://doi.org/10.1109/TKDE.2019.2912574

# Chapter 5
# Advanced Learning Rate-Free Latent Factor Analysis via P$^2$SO

## 5.1 Overview

Numerous entities are frequently encountered in various big data-related applications like wireless sensor networks [1–3], bioinformatics applications [4–9], social networks [10–13], user-service QoS [14, 15] and electronic commerce systems [16–18]. With the increasing number of involved entities, it becomes impossible to observe their whole interaction mapping. Hence, a resultant interaction mapping can be denoted by a high-dimensional and sparse (HiDS) matrix [19–23] with a few entries known (describing the observed interactions) while the most others unknown (describing the unobserved ones).

Although an HiDS matrices is extremely sparse, it involves a mountain of valuable information various patterns. In order to extract useful knowledge from an HiDS matrix, a pyramid of models have been proposed [24, 25]. Among them, a latent factor analysis (LFA) model stands out due to its high scalability and efficiency [23, 26–30]. An LFA model maps the rows and columns of an HiDS matrix into the same low-dimensional latent factor (LF) space, and builds the objective functions only based on the known values of the target matrix and desired LFs. Then it minimizes the objective function to train LFs and then predict the missing values in an HiDS matrix according to these obtained LFs.

When building an LFA model, a stochastic gradient descent (SGD) algorithm is very effective [31, 32]. However, the performance of an SGD-based LFA model is greatly influenced by its learning rate. An inappropriately large learning rate makes the model diverge while a too small one leads to slow model convergence. Especially, it is time-consuming when performing latent factor analysis on a huge HiDS matrix. Although many extended gradient descent algorithms have been proposed to address the issue of learning rate adaptation in an SGD algorithm, they drastically cost much more time in a single iteration than a standard SGD algorithm does.

Thus, we have the first research question of this chapter:

© The Author(s), under exclusive license to Springer Nature Singapore Pte Ltd. 2022
Y. Yuan, X. Luo, *Latent Factor Analysis for High-dimensional and Sparse Matrices*,
SpringerBriefs in Computer Science, https://doi.org/10.1007/978-981-19-6703-0_5

*RQ.* **1.** How to implement efficient learning rate adaptaion for an LFA model without additional computation burden?

Recent studies [33, 34] provide a probable answer to this question. For a general learning algorithm, a particle swarm optimization (PSO) algorithm can enable its hyper-parameter' adaptation due to its high efficiency and compatibility. However, a standard PSO algorithm commonly suffers from premature convergence when addressing a complex data analysis tasks [35, 36], i.e., performing LFA on an HiDS matrix is a bi-convex problem.

As unveiled by prior study [37–39], various PSO extensions have been proposed to address the premature issue of a standard PSO algorithm. However, such PSO extensions can alleviate premature convergence encountered in a standard PSO algorithm, they tend to perplex the evolution process of a standard PSO algorithm with more parameters or more complex searching strategies.

Hence, we have encountered the second research question of this chapter:

*RQ.* **2.** How to implement an extended PSO algorithm without premature convergence efficiently?

To address these issues, we first propose a novel position-transitional particle swarm optimization (P$^2$SO) algorithm by incorporate more dynamic information into the particle's evolution of a standard particle swarm optimization (PSO) algorithm, thereby avoiding accuracy loss caused by its premature convergence. Subsequently, we adopt P$^2$SO into the training process of an SGD-based LFA model for making learning rate adaptation with high efficiency, thereby achieving an an *Advanced Learning rate-free LFA* (AL$^2$FA) model. The contributions of this chapter are threefold:

(a) A P$^2$SO algorithm. It incorporates more dynamic information in a PSO algorithm, thereby improving the swarm searching ability;
(b) An AL$^2$FA model. It implements efficient learning rate adaptaion without prediction accuracy loss; and
(c) Detailed algorithm design and analysis of an AL$^2$FA model.

Experimental results on four HiDS matrices demonstrate that owing to the good searching ability and high computational efficiency of a P$^2$SO algorithm, an AL$^2$FA model's computational efficiency and prediction accuracy for missing data of an HiDS matrix are highly competitive.

The remainder of this chapter is organized as follows: Sect. 5.2 describes an SGD-based LFA model. Section 5.3 presents the proposed AL$^2$FA model in detail. Section 5.4 provides and analyses the empirical study results. Section 5.5 concludes the chapter.

## 5.2 An SGD-Based LFA Model

As shown in Sect. 1.2.2, the objective function of an LFA model is formulated by:

$$\varepsilon(X, Y) = \sum_{r_{m,n} \in \Lambda} \left( (r_{m,n} - \tilde{r}_{m,n})^2 + \lambda \|x_{m,\cdot}\|_2^2 + \lambda \|y_{n,\cdot}\|_2^2 \right)$$

$$= \sum_{r_{m,n} \in \Lambda} \left( (r_{m,n} - \tilde{r}_{m,n})^2 + \lambda \sum_{d=1}^{f} x_{m,d}^2 + \lambda \sum_{d=1}^{f} y_{n,d}^2 \right), \tag{5.1}$$

where $\tilde{r}_{m,n} = \sum_{d=1}^{f} x_{m,d} y_{n,d}$, $\lambda$ denotes the regularization coefficient, $x_{m,\cdot}$ is the $m$-th row vector in $X$, $y_{n,\cdot}$ is the $n$-th row vector and $Y$, $r_{m,n}$, $x_{m,d}$ and $y_{n,d}$ denote the single elements of $R$, $X$ and $Y$, $\|\cdot\|_2$ calculates the $L_2$ norm of the enclosed vector.

As indicated by prior research [23], SGD enjoys its fast convergence and ease of implementation when performing LFA on an HiDS matrix,

$$\operatorname*{argmin}_{X,Y} \varepsilon(X, Y) \overset{SGD}{\Rightarrow} \forall_{r_{m,n}} \in \Lambda, d \in \{1, 2, \cdots, f\} :$$

$$x_{m,d}^{\tau} \leftarrow x_{m,d}^{\tau-1} - \eta \frac{\partial \varepsilon_{m,n}^{\tau-1}}{\partial x_{m,d}^{\tau-1}}, y_{n,d}^{\sigma} \leftarrow y_{n,d}^{\sigma-1} - \eta \frac{\partial \varepsilon_{m,n}^{\sigma-1}}{\partial y_{n,d}^{\sigma-1}}; \tag{5.2}$$

where $\varepsilon_{m,n} = (r_{m,n} - \tilde{r}_{m,n})^2 + \lambda_X \|x_{m,\cdot}\|_2^2 + \lambda_Y \|y_{n,\cdot}\|_2^2$ is the instant error corresponding to the training instance $r_{m,n} \in \Lambda$, and $\tau$ and $(\tau - 1)$ denote the current and last update points for $x_{m,d}$, $\sigma$ and $\sigma - 1$ denote the current and last update points for $y_{n,d}$, respectively. Based on (5.1), we achieve the following expressions:

$$\frac{\partial \varepsilon_{m,n}^{\tau-1}}{\partial x_{m,d}^{\tau-1}} = -err_{m,n}^{\tau-1} \cdot y_{n,d}^{\sigma-1} + \lambda x_{m,d}^{\tau-1},$$

$$\frac{\partial \varepsilon_{m,n}^{\sigma-1}}{\partial y_{n,d}^{\sigma-1}} = -err_{m,n}^{\sigma-1} \cdot x_{m,d}^{\tau-1} + \lambda y_{n,d}^{\sigma-1}; \tag{5.3}$$

where $err_{m,n} = r_{m,n} - \overset{\ddot{A}}{r}_{m,n}$, and substituting (5.3) into (5.2), we achieve the following update rules of LF matrices $X$ and $Y$ as follows:

$$\operatorname*{argmin}_{X,Y} \varepsilon(X, Y) \overset{SGD}{\Rightarrow} \forall_{r_{m,n}} \in \Lambda, d \in \{1, 2, \cdots, f\} :$$

$$\begin{cases} x_{m,d}^{\tau} \leftarrow x_{m,d}^{\tau-1} + \eta \left( err_{m,n}^{\tau-1} \cdot y_{n,d}^{\sigma-1} - \lambda x_{m,d}^{\tau-1} \right), \\ y_{n,d}^{\sigma} \leftarrow y_{n,d}^{\sigma-1} + \eta \left( err_{m,n}^{\sigma-1} \cdot x_{m,d}^{\tau-1} - \lambda y_{n,d}^{\sigma-1} \right). \end{cases} \tag{5.4}$$

As shown in (5.4), an SGD-based LFA model can converge fast with appropriate chosen values of $\eta$ [40, 41], which actually arises from the Robbins-Siegmund theorem [38].

## 5.3   The Proposed AL²FA Model

### 5.3.1   A P²SO Algorithm

As shown in Sect. 1.2.3, we can find that each particle of a standard PSO algorithm depends on three learning factors: (a) latest evolution velocity $v_j(t-1)$, (b) latest position $s_j(t-1)$ to its latest locally optimal position $pb_j(t-1)$, and (c) latest position $s_j(t-1)$ to the latest globally optimal position $gb(t-1)$.

Based on (1.3), we can find that randomness is injected into each particle's evolution process to enhance its searching ability of each particle. However, each particle evolves based on the 'statistic' positions, i.e., $s_j(t-1), pb_j(t-1)$ and $gb(t-1)$, making the whole swarm be stacked by some local solution. From this point of view, we propose to incorporate more 'dynamic' information into the evolution rule of each particle. Hence, the evolution rule of the $j$-th particle at the $t$-th iteration can be reformulated as:

\scale90%
$$
\begin{cases}
v_j(t) = wv_j(t-1) + c_1r_1\big(pb_j(t-1) - s_j(t-1)\big) + c_2r_2\big(gb(t-1) - s_j(t-1)\big) \\
\quad + \rho(c_1r_1 + c_2r_2)\big(s_j(t-1) - v_j(t-1)\big), \\
s_j(t) = s_j(t-1) + v_j(t);
\end{cases}
$$

$$(5.5)$$

where $\rho$ denotes a coefficient in the range of [0, 1]. Next, we divide it into three cases:

1. *Case $\rho = 0$*

We clearly find that (5.5) is equivalent to a standard PSO algorithm by substituting $\rho = 0$. As shown in Fig. 5.1, the evolution of each particle in a standard is mostly

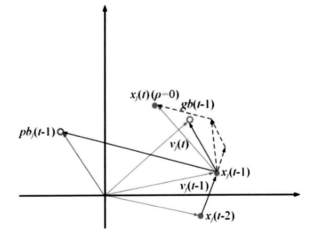

**Fig. 5.1** Evolution direction with the case $\rho = 0$ (standard PSO)

**Fig. 5.2** Evolution
direction with the case $\rho = 1$

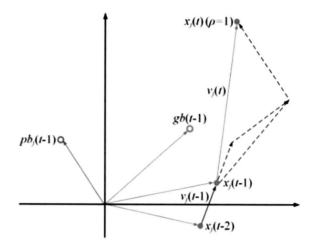

decided by the adjusting velocities, i.e., $(pb_j(t-1) - s_j(t-1))$ and $(gb(t-1) - s_j(t-1))$. Hence, each particle tends to approach its local optimal position and global optimal position of the swarm very quickly, making the whole swarm suffer premature convergence.

2. *Case $\rho = 1$*

If we substitute $\rho = 1$ into (5.5), another evolution rule is achieved:

$$\begin{cases} v_j(t) = wv_j(t-1) + c_1 r_1 \big(pb_j(t-1)\big) - c_1 r_1 v_j(t-1) \\ \quad + c_2 r_2 \big(gb(t-1) - c_2 r_2 v_j(t-1)\big), \\ s_j(t) = s_j(t-1) + v_j(t). \end{cases} \tag{5.6}$$

Note that in (5.6) each particle's evolution depends on the following three terms: (a) the latest evolution velocity, i.e., $v_j(t-1)$, (b) the adjusting velocity decided by the locally optimal position and the latest evolution velocity, i.e., $(pb_j(t-1) - 0)$ and $v_j(t-1)$, and (c) the adjusting velocity decided by the globally optimal position and the latest evolution velocity, i.e., $(gb(1) - 0)$ and $v_j(t-1)$. As shown in Fig. 5.2, more dynamic information relying on $v_j(t-1)$ is injected into the particle evolution, therefore making the whole swarm more active, thereby avoiding premature convergence.

3. *Case $0 < \rho < 1$*

When $\rho$ lies in the scale of (0, 1), (5.6) can be transformed into:

**Fig. 5.3** Evolution
direction with the case
$0 < \rho < 1$

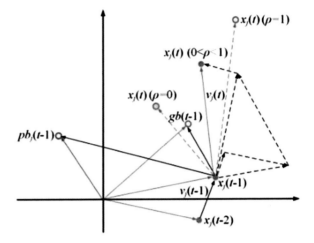

$$\begin{cases} v_j(t) = wv_j(t-1) + c_1r_1\big(pb_j(t-1) - s_j(t-1)\big) + c_2r_2\big(gb(t-1) - s_j(t-1)\big) \\ \quad + \rho c_1 r_1\big(s_j(t-1) - v_j(t-1)\big) + \rho c_2 r_2\big(s_j(t-1) - v_j(t-1)\big), \\ s_j(t) = s_j(t-1) + v_j(t). \end{cases}$$

$$(5.7)$$

From (5.7), we see that with $0 < \rho < 1$ we arrive at an intermediate state between two extreme cases of $\rho = 0$ and $\rho = 1$. Note that the evolution velocity does not only depend on the adjusting velocities decided by the static positions, but also depend on the dynamic evolution velocities as analyzed in the previous section. The evolution process is depicted in Fig. 5.3.

Based on the above inferences, we achieve a P²SO algorithm. Next, we aim at implementing self-adaptation of $\eta$ in an AL²FA model via a P²SO algorithm.

## 5.3.2  Learning Rate Adaptation via P²SO

Firstly, we build a swarm consisting of $q$ particles, where the $j$-th particle is the learning rate $\eta_j$ for the same group of LFs. With a P²SO algorithm, $\eta_j$'s evolution scheme is given by:

$$\begin{cases} v_j(t) = wv_j(t-1) + c_1r_1\big(pb_j(t-1) - \eta_j(t-1)\big) + c_2r_2\big(gb(t-1) - \eta_j(t-1)\big) \\ \quad + \rho(c_1r_1 + c_2r_2)\big(\eta_j(t-1) - v_j(t-1)\big), \\ \eta_j(t) = \eta_j(t-1) + v_j(t). \end{cases}$$

$$(5.8)$$

The position and velocity of each particle must be constrained in a certain range:

$$v_j(t) = \begin{cases} \check{v} & v_j(t) > \check{v} \\ \check{v}\square & v\square v_j(t) < v\square \end{cases}, \quad \eta_j(t) = \begin{cases} \check{s} & s_j(t) > \check{s} \\ s\square & s\square s_j(t) < s\square \end{cases}. \tag{5.9}$$

where $\check{v}$ and $v\square$ denote the upper and lower bounds for $v$, $\check{s}$ and $s\square$ denote the upper and lower bounds for $\eta$, respectively. We commonly have $\check{v} = -\square v = 1d$, $\check{s} = 2^{-8}$, and $s\square = 2^{-12}$.

For better acquiring LFs from an HiDS matrix, we adopt the following two fitness functions for the $j$-th particle:

$$F_{1(j)} = \sqrt{\left( \sum_{r_{m,n} \in \Omega} \left( r_{m,n} - \tilde{r}_{(j)m,n} \right)^2 \right) / |\Omega|},$$

$$F_{2(j)} = \left( \sum_{r_{m,n} \in \Omega} \left| r_{m,n} - \tilde{r}_{(j)m,n} \right|_{abs} \right) / |\Omega|; \tag{5.10}$$

where $|\cdot|_{abs}$ calculates the absolute value of a given value, $\Omega$ represents the validation set and is disjoint with $\Lambda$, and $\tilde{r}_{(j)m,n}$ represents the estimation value regarding to the known element $r_{m,n} \in \Omega$ generated by the $j$-th particle, respectively. To be shown later, an A²LFA model adopts $F_{1(j)}$ or $F_{2(j)}$ according to the performance evaluation metrics.

Note that $\forall j \in \{1, \ldots, q\}$, $\eta_j$ is linked with the same group of LF matrices, i.e., $X$ and $Y$, which are trained by an SGD solver. Thus, its $t$-th iteration actually consists of $q$ sub-iterations, where $X$ and $Y$ are updated in the $j$-th sub-iteration as follows:

$$\underset{X,Y}{\operatorname{argmin}} \, \varepsilon(X, Y) \Rightarrow \forall_{r_{m,n}} \in \Lambda, d \in \{1, 2, \cdots, f\}:$$

$$\begin{cases} x_{(j)m,d}^{\tau} \leftarrow x_{(j)m,d}^{\tau-1} + \eta_j^{t-1} \cdot \left( err_{(j)m,n}^{\tau-1} \cdot y_{(j)n,d}^{\sigma-1} \cdot \lambda x_{(j)m,d}^{\tau-1} \right), \\ y_{(j)n,d}^{\sigma} \leftarrow y_{(j)n,d}^{\sigma-1} + \eta_j^{\sigma-1} \cdot \left( err_{(j)m,n}^{\sigma-1} \cdot x_{(j)m,d}^{\tau-1} \cdot \lambda y_{(j)n,d}^{\sigma-1} \right). \end{cases} \tag{5.11}$$

where the footnote $(j)$ on $x_{m,d}$, $y_{n,d}$ and $err_{m,n}$ denotes that their current updates are implemented with the $j$-th particle.

### 5.3.3 Algorithm Design and Analysis

Based on the inferences, we design the algorithm AL²FA. From each step in Algorithm AL²FA, we see that its computational complexity is:

$$T = \Theta\big(((|M| + |N|) \times f + q + 2q \times D + t \times (q \times |\Lambda| \times 3f + 14q))\big)$$
$$\approx \Theta(|\Lambda| \times t \times q \times f) \tag{5.12}$$

Note that in (5.12) with $|\Lambda| \ll \max\{|M|, |N|\}$, we can omit the lower-order-terms to ensure the result reasonably. Based on the above inferences, we see that the AL$^2$FA algorithm is highly efficient in computation.

| ALGORITHM AL$^2$FA | |
|---|---|
| **Input:** $\Lambda$, $U$, $I$, $f$, $\lambda_P$, $\lambda_Q$, $S$, $D$, $C$, $c_1$, $c_2$ | |
| **Operation** | **Cost** |
| **Initialize** $X^{|M| \times f}$, $Y^{|N| \times f}$ | $\Theta(|M| +$ $ N|) \times f$ |
| **Initialize** $r_1$, $r_2$, $w$, $gb$ | $\Theta(1)$ |
| **Initialize** $V^{S \times D}$, $H^{S \times D}$ | $\Theta(S \times D)$ |
| **Initialize** $pb^S$ | $\Theta(S)$ |
| **Initialize** $v_{\max}$, $v_{\min}$, $x_{\max}$, $x_{\min}$, $t = 1$, $T = $ Maximum Round Count | $\Theta(1)$ |
| **while not** converge **and** $t \le T$ **do** | $\times T$ |
|  **for** each $j \in S$ | $\times S$ |
|   **for** each $z_{u,i} \in \Lambda$ | $\times |\Lambda|$ |
|    $\hat{z}\{z\}_{u,i} = p_u q_i^T$ | $\Theta(f)$ |
|    $err = z_{u,i} - \hat{z}\{z\}_{u,i}$ | $\Theta(1)$ |
|    **for** $d = 1$ **to** $f$ | $\times f$ |
|     $x_{m,d} = x_{m,d} + \eta_j \times (err \times y_{n,d} - \lambda \times x_{m,d})$, $y_{n,d} = y_{n,d} + \eta_j \times (err \times x_{m,d} - \lambda \times y_{n,d})$ | $\Theta(1)$ |
|    **end for** | – |
|    $F_j = fitness$ | $\Theta(1)$ |
|   **end for** | – |
|  **for** each $j \in S$ | $\times S$ |
|   **if** $F < pb_j$ **then** $pb_j = F$ | $\Theta(1)$ |
|   **if** $F < gb$ **then** $gb = F$ | $\Theta(1)$ |
|  **end for** | – |
|  **for** each $j \in S$ | $\times S$ |
|   $v_j = wv_j + c_1 r_1(pb_j - ((1 - \rho)\eta_j + \rho v_j)) + c_2 r_2(gb - ((1 - \rho)\eta_j + \rho v_j))$ | $\Theta(1)$ |
|   **if** $v_j > v_{\max}$ **then** $v_j = v_{\max}$ **else** $v_j = v_{\min}$ | $\Theta(1)$ |
|   $\eta_j = \eta_j + v_j$ | $\Theta(1)$ |
|   **if** $\eta_j > \eta_{\max}$ **then** $\eta_j = \eta_{\max}$ **else** $\eta_j = \eta_{\min}$ | $\Theta(1)$ |
|  **end for** | – |
|  $t = t + 1$ | $\Theta(1)$ |
| **end while** | – |
| **Output:** $P$, $Q$ | |

## 5.4 Experimental Results and Analysis

### 5.4.1 General Settings

**Evaluation protocol** For a tested model, root mean squared error (RMSE) and mean absolute error (MAE) are frequently used to validate a tested model's prediction accuracy for missing data of an HiDS matrix [1–3, 23–26, 42]:

$$
\text{RMSE} = \sqrt{\left(\sum_{r_{m,n} \in \Phi} (r_{m,n} - \tilde{r}_{m,n})^2\right) / |\Phi|},
$$

$$
\text{MAE} = \frac{\left(\sum_{r_{m,n} \in \Phi} |r_{m,n} - \tilde{r}_{m,n}|_{abs}\right)}{|\Phi|};
\tag{5.13}
$$

where $\ddot{r}_{m,n}$ denotes the generated prediction for the testing instance $r_{m,n} \in \Phi$, which represents the missing value at the $m$-th row and $n$-th column of a target HiDS matrix. $|\cdot|$ represents the cardinality of a given set. $|\Phi|$ denotes the size of the validation dataset $\Omega$.

**Datasets** Four HiDS matrices adopted in our experiments. Their details are recorded in Table 5.1.

Note that each dataset is randomly split into ten disjoint subsets for implementing 70%–10%–20% train-validation-test settings. More specifically, on each dataset, we adopt seven subsets as a training set to train a model, one as a validating set to monitor the training process for making the model achieve its optimal outputs, and the last two as the testing set for verifying the performance of each tested model. This process is sequentially repeated ten times to acquire the final results. The termination condition is uniform for all involved models, i.e., the iteration threshold is 1000, and error threshold is $10^{-5}$, and a tested model's training process terminates if either threshold is met.

**Model Settings.** To obtain a fair and objective comparison, we adopt the following measures settings:

(a) LF matrices $X$ and $Y$ are initialized randomly generated in the range of $[-0.05, 0.05]$ for eliminating the performance bias;

**Table 5.1** Datasets details

| No. | Name | Row | Column | Known entries | Density (%) |
|-----|------|-----|--------|---------------|-------------|
| D1 | MovieLens 10M | 10,681 | 71,567 | 10,000,054 | 1.31 |
| D3 | Epinion | 120,492 | 775,760 | 13,668,320 | 0.015 |
| D3 | Douban | 58,541 | 129,490 | 16,830,839 | 0.22 |
| D4 | Flixter | 48,794 | 147,612 | 8,196,077 | 0.11 |

(b) LF dimension $f = 20$ to balance the representative learning ability and compu-
    tational efficiency;
(c) We adopt the same empirical value for regularization coefficients on $X$ and $Y$,
    i.e., setting $\lambda = 0.03$ uniformly; and
(d) In PSO and P²SO, we choose the same settings for the hyper parameters: $w$, $r_1$
    and $r_2 \in (0,1)$ and is generated by a uniform distribution, $S = 10$, $c_1 = c_2 = 2$, $v \in$
    $[-1,1]$, and $\forall j \in \{1, \ldots, q\}$: $\eta_j \in [2^{-12}, 2^{-8}]$.

### 5.4.2  Effect of $\rho$

As discussed in Sect. 5.3.1, a P²SO algorithm tunes the static and dynamic adjusting
velocities through the value of $\rho$. Hence, the value of $\rho$ affects AL²FA's performance. In this part, we conduct parameter sensitivity tests, which are given in
Figs. 5.4 and 5.5. From them, we have the following findings:

(a) AL²FA's prediction error decreases as $\rho$ increases. For instance, as depicted in
    Fig. 5.4a, AL²FA's achieves the lowest RMSE at 0.7858 with $\rho = 1$ and the
    highest RMSE at 0.7993, respectively. The gap between them reaches 1.69%.
    Similar situations are found on other testing cases. The main reason is that P²SO

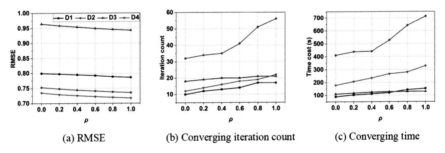

|  (a) RMSE  |  (b) Converging iteration count  |  (c) Converging time  |

**Fig. 5.4** AL²FA's performance corresponding to RMSE as $\rho$ varies. All panels share the legend in
panel (**a**)

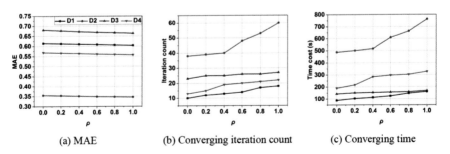

|  (a) MAE  |  (b) Converging iteration count  |  (c) Converging time  |

**Fig. 5.5** AL²FA's performance corresponding to MAE as $\rho$ varies. All panels share the legend in
panel (**a**)

becomes PSO with $\rho = 0$. The standard PSO depends on 'static' adjusting velocities solely depending on the static positions connected with each particle. In contrast, $P^2SO$ incorporates more 'dynamic' information into each particle's evolution process, which is beneficial for $AL^2FA$'s prediction accuracy.

(b) $AL^2FA$'s iterations increases as $\rho$ increases. For instance, as shown in Fig. 5.4b, on D1, $AL^2FA$'s iterations is 10 and 17 as $\rho = 0$ and 1, respectively. Similar situations are observed on other testing cases. Note that $P^2SO$ prevents premature convergence by incorporating more 'dynamic' information. The main reason is that $AL^2FA$ consumes more iterations as the value of $\rho$ increases. Moreover, $AL^2FA$'s total time cost also increases as $\rho$ increases.

(c) $AL^2FA$'s performance relies heavily on the value of $\rho$. Note that when it is small, $AL^2FA$ converges with fewer iterations and less time cost but suffering a loss of prediction accuracy. With it increases, the iterations and total time cost increases, but it obtains an accuracy gain. However, to be shown next, owing to its self-adaptation of learning rate, $AL^2FA$ is able to achieve very competitive prediction accuracy for missing data, as well as very high computational efficiency when compared with a SGD-based LFA model.

### 5.4.3 Comparison Results

In this part of the experiments, we compare the proposed $AL^2FA$ model with several state-of-the-art models on estimation accuracy and computational efficiency for missing data of an HiDS matrix. The details of all the compared models are summarizes in Table 5.2 as followings:

Note that the learning rate of M1 is not self-adaptive. Hence, we have tuned its learning rate via grid-search on each dataset to make it achieve the highest prediction accuracy. Tables 5.3, 5.4, 5.5, and 5.6 summarize the comparison results. From them, we have the following findings:

**Table 5.2** Details of compared models

| No. | Model |
|---|---|
| M1 | An SGD-based LFA model [23] |
| M2 | An $AL^2FA$ model with $\rho = 0$ |
| M3 | An Adam-based LFA model [23] |
| M4 | An $AL^2FA$ model proposed in this chapter |

**Table 5.3** Lowest RMSE of M1–M4

| Dataset | Lowest RMSE/converging iteration count | | | |
|---|---|---|---|---|
| | M1 | M2 | M3 | M4 |
| D1 | 0.7872/1000 | 0.7993/**10** | 0.7901/113 | **0.7858**/17 |
| D2 | 0.7358/1000 | 0.7531/**32** | **0.7342**/419 | 0.7354/56 |
| D3 | 0.9433/657 | 0.9641/**18** | 0.9464/542 | **0.9429**/21 |
| D4 | 0.7176/764 | 0.7352/**12** | 0.7213/151 | **0.7173**/22 |

**Table 5.4** Time cost corresponding to Table 5.3 (s)

| Dataset | Time cost per iteration/total time cost | | | |
| --- | --- | --- | --- | --- |
| | M1 | M2 | M3 | M4 |
| D1 | **0.874**/874 | 8.859/**89** | 7.885/891 | 9.019/153 |
| D2 | **1.151**/1151 | 12.838/**411** | 10.248/4294 | 12.751/714 |
| D3 | **0.644**/423 | 6.065/**109** | 6.141/3328 | 6.185/130 |
| D4 | **1.552**/1186 | 14.851/**178** | 13.231/1998 | 14.948/329 |

**Table 5.5** Lowest MAE of M1–M4

| Dataset | Lowest MAE/converging iteration count | | | |
| --- | --- | --- | --- | --- |
| | M1 | M2 | M3 | M4 |
| D1 | 0.6069/1000 | 0.6147/**10** | 0.6087/116 | **0.6040**/18 |
| D2 | 0.3467/1000 | 0.3552/**38** | **0.3451**/426 | 0.3462/60 |
| D3 | 0.6653/703 | 0.6809/**23** | 0.6661/552 | **0.6641**/27 |
| D4 | 0.5578/771 | 0.5688/**13** | 0.5582/156 | **0.5574**/22 |

**Table 5.6** Time cost corresponding to Table 5.5 (s)

| Dataset | Time cost per iteration/total time cost | | | |
| --- | --- | --- | --- | --- |
| | M1 | M2 | M3 | M4 |
| D1 | **0.869**/869 | 8.859/**89** | 7.882/914 | 8.896/160 |
| D2 | **1.152**/1152 | 12.827/**487** | 10.239/4331 | 12.716/763 |
| D3 | **0.641**/451 | 6.188/**142** | 6.146/3393 | 6.245/169 |
| D4 | **1.547**/1193 | 14.583/**190** | 13.228/2064 | 14.875/327 |

(a) AL²FA implements efficient self-adaptation of learning rate without loss of prediction accuracy due to its incorporation of P²SO. For instance, as shown in Table 5.3, on D3, M4 achieves the lowest RMSE at 0.9429, which is 0.042% lower than the RMSE at 0.9433 by M1. Similar situations are also found on other testing cases. Although the accuracy gain is slight, such phenomenon is very impressive. This is because M1's learning rate is carefully tuned to enable it to achieve the highest prediction accuracy for missing data on each testing case. Compared M1 and M2, we can find that the prediction error of M2 is always higher than M1, which is considered as its compromise between the adaptive learning rate and prediction accuracy for missing data. Moreover, M3 is also a learning rate-adaptive algorithm, but it still suffers prediction accuracy loss except on D2. The same situations are found in Table 5.5.

(b) AL²FA's convergence rate is fast. For instance, as summarized in Table 5.3, M4 takes 17 iterations to converge in RMSE on D1, while M1 takes 1000. M4's converging iteration count is only 1.7% that of M1's. Similar situations are also encountered on other testing cases. However, M4's iterations is higher than that of M2 on all the testing cases. The main reason is that P²SO incorporates more 'dynamic' information to prevent premature convergence.

(c) AL²FA's computational efficiency is high. Due to its fast convergence rate, M4 takes much less total time than M1 does. For instance, as recorded in Table 5.4, on D1, M4 consumes 153 s to achieve the lowest RMSE, only 17.51% of 874 s by M1 does. The same situations are also found on the other testing cases. Moreover, as shown in Table 5.4, M2 consumes less total time than M4 does, but it suffers significant accuracy loss. On the other hand, M3 makes compromise of time cost for its self-adaptive learning rate. Its total time cost is even significantly higher than M1. The same situations are found in Table 5.6.

### 5.4.4 Summary

Based on the experimental results, we summarize that:

(a) AL²FA will not suffer premature convergence by incorporating more 'dynamic' information; and
(b) AL²FA implements self-adaptation of its learning rate with high prediction accuracy for missing data of an HiDS matrix along with low time cost.

## 5.5 Conclusions

This chapter proposes an AL²FA model, which implements efficient self-adaptation of learning rate without loss of prediction accuracy due to its incorporation of P²SO. Empirical results show that with such design, AL²FA achieves high prediction accuracy and computational efficiency for missing data of an HiDS matrix with learning rate adaptation. Moreover, it also shows great potential for other optimization problems in industrial applications, e.g., the optimization of robot control [43–52].

In the future, we aim to address the following issues:

(a) Many improved PSO algorithms have shown their impressive ability in addressing general optimization problems. Can we achieve further performance gain by incorporating more 'dynamic' information into them? This question remains open; and
(b) Do other intelligent optimization algorithms have the potential to build an LFA models with learning rate adaptation? For instance, a beetle antennae search algorithm [45] can solve an optimization problem in way of evolution conveniently, yet it commonly leads to accuracy loss. A brain storm optimization algorithm enables a learning model's high representation learning ability [53, 54], but it relies on an inner clustering process which can yield high computational complexity. It is highly interesting to implement hyper parameter-free LFA models based on their principles. However, great efforts are further desired to achieve so.

# References

1. Michaelides, C., Pavlidou, F.N.: Mutual aid among sensors: an emergency function for sensor networks. IEEE Sens. Lett. **4**(9), 1–4 (2020)
2. Chaudhry, R., Kumar, N.: A multi-objective meta-heuristic solution for green computing in software-defined wireless sensor networks. IEEE Trans. Green Commun. Netw. **6**(2), 1231–1241 (2022)
3. Omeke, K.G., et al.: DEKCS: a dynamic clustering protocol to prolong underwater sensor networks. IEEE Sens. J. **21**(7), 9457–9464 (2021)
4. Luo, X., You, Z., Zhou, M., Li, S., Leung, H., Xia, Y., Zhu, Q.: A highly efficient approach to protein interactome mapping based on collaborative filtering framework. Sci. Rep. **5**, 7702 (2015)
5. Hofree, M., Shen, J.P., Carter, H., Gross, A., Ideker, T.: Network-based stratification of tumor mutations. Nat. Methods. **10**(11), 1108–1115 (2013)
6. You, Z.H., Lei, Y.K., Gui, J., Huang, D.S., Zhou, X.B.: Using manifold embedding for assessing and predicting protein interactions from high-throughput experimental data. Bioinformatics. **26**(21), 2744–2751 (2010)
7. Hu, L., Yang, S.C., Luo, X., Yuan, H.Q., Zhou, M.C.: A distributed framework for large-scale protein-protein interaction data analysis and prediction using MapReduce. IEEE/CAA J. Autom. Sin. **9**(1), 160–172 (2022). https://doi.org/10.1109/JAS.2021.1004198
8. Hu, L., Yuan, X.H., Liu, X., Xiong, S.W., Luo, X.: Efficiently detecting protein complexes from protein interaction networks via alternating direction method of multipliers. IEEE/ACM Trans. Comput. Biol. Bioinform. **16**(6), 1922–1935 (2019)
9. You, Z.H., Zhou, M.C., Luo, X., Li, S.: Highly efficient framework for predicting interactions between proteins. IEEE Trans. Cybern. **64**(6), 4710–4720 (2017)
10. Cao, X., Wang, X., Jin, D., Cao, Y., He, D.: Identifying overlapping communities as well as hubs and outliers via nonnegative matrix factorization. Sci. Rep. **3**, 2993 (2013)
11. Liu, S.X., Hu, X.J., Wang, S.H., Zhang, Y.D., Fang, X.W., Jiang, C.Q.: Mixing patterns in social trust networks: a social identity theory perspective. IEEE Trans. Comput. Soc. Syst. **8**(5), 1249–1261 (2021)
12. Whitaker, R.M., et al.: The coevolution of social networks and cognitive dissonance. IEEE Trans. Comput. Soc. Syst. **9**(2), 376–393 (2022)
13. Luo, X., Zhou, M.-C., Wang, Z.-D., Xia, Y.-N., Zhu, Q.-S.: An effective QoS estimating scheme via alternating direction method-based matrix factorization. IEEE Trans. Serv. Comput. **12**(4), 503–518 (2019)
14. Luo, X., Chen, M.Z., Wu, H., Liu, Z.G., Yuan, H.Q., Zhou, M.C.: Adjusting learning depth in non-negative latent factorization of tensors for accurately modeling temporal patterns in dynamic QoS data. IEEE Trans. Autom. Sci. Eng. **18**(4), 2142–2155 (2021). https://doi.org/10.1109/TASE.2020.3040400
15. Wu, D., Luo, X., Shang, M.S., He, Y., Wang, G.Y., Wu, X.D.: A data-characteristic-aware latent factor model for web services QoS prediction. IEEE Trans. Knowl. Data Eng. **34**(6), 2525–2538 (2022). https://doi.org/10.1109/TKDE.2020.3014302
16. Li, Y., Cao, B., Xu, L., Yin, J.W., Deng, S.G., Yin, Y.Y., Wu, Z.H.: An efficient recommendation method for improving business process modeling. IEEE Trans. Ind. Inform. **10**(1), 502–513 (2013)
17. Zhang, W., Zhang, X., Chen, D.: Causal neural fuzzy inference modeling of missing data in implicit recommendation system. Knowl.-Based Syst. **222**(11), 106678 (2021)
18. Da'U, A., Salim, N., Idris, R.: An adaptive deep learning method for item recommendation system. Knowl.-Based Syst. **213**(8), 106681 (2021)
19. Luo, X., Zhong, Y.R., Wang, Z.D., Li, M.Z.: An alternating-direction-method of multipliers-incorporated approach to symmetric non-negative latent factor analysis. IEEE Trans. Neural Netw. Learn. Syst. https://doi.org/10.1109/TNNLS.2021.3125774

20. Luo, X., Zhou, Y., Liu, Z.G., Hu, L., Zhou, M.C.: Generalized Nesterov's acceleration-incorporated, non-negative and adaptive latent factor analysis. IEEE Trans. Serv. Comput. https://doi.org/10.1109/TSC.2021.3069108
21. Shi, X.Y., He, Q., Luo, X., Bai, Y.N., Shang, M.S.: Large-scale and scalable latent factor analysis via distributed alternative stochastic gradient descent for recommender systems. IEEE Trans. Big Data. **8**(2), 420–431 (2022). https://doi.org/10.1109/TBDATA.2020.2973141
22. Wu, D., He, Y., Luo, X., Zhou, M.C.: A latent factor analysis-based approach to online sparse streaming feature selection. IEEE Trans. Syst. Man Cybern. Syst. https://doi.org/10.1109/TSMC.2021.3096065
23. Luo, X., Zhou, M.C.: Effects of extended stochastic gradient descent algorithms on improving latent factor-based recommender systems. IEEE Robot. Autom. Lett. **4**(2), 618–624 (2019)
24. Chu, W., Ghahramani, Z.: Probabilistic models for incomplete multi-dimensional arrays. In: Proc. of the 12th Int. Conf. on Artificial Intelligence and Statistics, Clearwater Beach, FL, pp. 89–96 (Apr 2009)
25. Chatzis, S.: Nonparametric Bayesian multitask collaborative filtering. In: Proc. of the 22nd ACM Int. Conf. on Information and Knowledge Management, San Francisco, CA, pp. 2149–2158 (Oct 2013)
26. Wu, J., Chen, L., Feng, Y.-P., Zheng, Z.-B., Zhou, M.-C., Wu, Z.-H.: Predicting quality of service for selection by neighborhood-based collaborative filtering. IEEE Trans. Syst. Man Cybern. Syst. **43**(2), 428–439 (2013)
27. Pan, J.-J., Pan, S.-J., Jie, Y., Ni, L.-M., Yang, Q.: Tracking mobile users in wireless networks via semi-supervised colocalization. IEEE Trans. Pattern Anal. Mach. Intell. **34**(3), 587–600 (2012)
28. Luo, X., Zhou, M.C., Li, S., You, Z.H., Xia, Y.N., Zhu, Q.S.: A non-negative latent factor model for large-scale sparse matrices in recommender systems via alternating direction method. IEEE Trans. Neural Netw. Learn. Syst. **27**(3), 524–537 (2016)
29. Luo, X., Zhou, M.C., Li, S., You, Z.H., Xia, Y.N., Zhu, Q.S., Leung, H.: An efficient second-order approach to factorizing sparse matrices in recommender systems. IEEE Trans. Ind. Inform. **11**(4), 946–956 (2015)
30. Yunxiao, C., Xiaoou, L., Siliang, Z.: Structured latent factor analysis for large-scale data: identifiability, estimability, and their implications. J. Am. Stat. Assoc. **115**, 1756–1770 (2020). https://doi.org/10.1080/01621459.2019.1635485
31. Qing, L., Diwen, X., Mingsheng, S.: Adjusted stochastic gradient descent for latent factor analysis. Inf. Sci. **588**, 196–213 (2022). https://doi.org/10.1016/J.INS.2021.12.065
32. Wu, H., Luo, X., Zhou, M.C., Rawa, M.J., Sedraoui, K., Albeshri, A.: A PID-incorporated latent factorization of tensors approach to dynamically weighted directed network analysis. IEEE/CAA J. Autom. Sin. **9**(3), 533–546. https://doi.org/10.1109/JAS.2021.1004308
33. Hsieh, S., Sun, T., Lin, C., Liu, C.: Effective learning rate adjustment of blind source separation based on an improved particle swarm optimizer. IEEE Trans. Evol. Comput. **12**(2), 242–251 (2008)
34. Zhan, Z.-H., Zhang, J., Li, Y., Chung, H.: Adaptive particle swarm optimization. IEEE Trans. Syst. Man Cybern. B Cybern. **39**(6), 1362–1381 (2009)
35. Yang, T., Wang, Z.-D., Fang, J.-A.: Parameters identification of unknown delayed genetic regulatory networks by a switching particle swarm optimization algorithm. Expert Syst. Appl. **38**(3), 2523–2535 (2011)
36. Wu, D.: Cloud computing task scheduling policy based on improved particle swarm optimization. In: 2018 Int. Conf. on Virtual Reality and Intelligent Systems (ICVRIS). https://doi.org/10.1109/ICVRIS.2018.00032
37. Behnamian, J., Ghomi, S.-M.: Development of a PSO-SA hybrid metaheuristic for a new comprehensive regression model to time-series forecasting. Expert Syst. Appl. **37**(2), 974–984 (2010)

38. Li, Y.-H., Zhan, Z.-H., Lin, S.-J., Zhang, J., Luo, X.-N.: Competitive and cooperative particle swarm optimization with information sharing mechanism for global optimization problems. Inf. Sci. **293**, 370–382 (2015)
39. Xia, X.-W., Ling, G., Zhan, Z.-H.: A multi-swarm particle swarm optimization algorithm based on dynamical topology and purposeful detecting. Appl. Soft Comput. **67**, 126–140 (2018)
40. Kiwiel, K.C.: Convergence and efficiency of subgradient methods for quasiconvex minimization. Math. Program. **90**(1), 1–25 (2001)
41. Krzysztof, C.K.: Convergence of approximate and incremental subgradient methods for convex optimization. SIAM J. Optim. **14**(3), 807–840 (2004)
42. Wu, H., Luo, X., Zhou, M.C.: Advancing non-negative latent factorization of tensors with diversified regularizations. IEEE Trans. Serv. Comput. **15**(3), 1334–1344 (2022). https://doi.org/10.1109/TSC.2020.2988760
43. Xie, Z.T., Jin, L., Luo, X., Hu, B., Li, S.: An acceleration-level data-driven repetitive motion planning scheme for kinematic control of robots with unknown structure. IEEE Trans. Syst. Man Cybern. Syst. **52**(9), 5679–5691 (2022). https://doi.org/10.1109/TSMC.2021.3129794
44. Chen, D.C., Li, S., Wu, Q., Luo, X.: New disturbance rejection constraint for redundant robot manipulators: an optimization perspective. IEEE Trans. Ind. Inform. **16**(4), 2221–2232 (2020)
45. Jin, L., Li, S., Luo, X., Li, Y.M., Qin, B.: Neural dynamics for cooperative control of redundant robot manipulators. IEEE Trans. Ind. Inform. **14**(9), 3812–3821 (2018)
46. Qi, Y., Jin, L., Luo, X., Zhou, M.C.: Recurrent neural dynamics models for perturbed nonstationary quadratic programs: a control-theoretical perspective. IEEE Trans. Neural Netw. Learn. Syst. **33**(3), 1216–1227 (2022). https://doi.org/10.1109/TNNLS.2020.3041364
47. Luo, X., Wu, H., Wang, Z., Wang, J.J., Meng, D.Y.: A novel approach to large-scale dynamically weighted directed network representation. IEEE Trans. Pattern Anal. Mach. Intell. https://doi.org/10.1109/TPAMI.2021.3132503
48. Khan, A.H., Li, S., Luo, X.: Obstacle avoidance and tracking control of redundant robotic manipulator: an RNN based metaheuristic approach. IEEE Trans. Ind. Inform. **16**(7), 4670–4680 (2020)
49. Xiao, X., Ma, Y., Xia, Y., Zhou, M., Luo, X., Wang, X., Fu, X., Wei, W., Jiang, N.: Novel workload-aware approach to mobile user reallocation in crowded mobile edge computing environment. IEEE Trans. Intell. Transport. Syst. **23**(7), 8846–8856 (2022). https://doi.org/10.1109/TITS.2021.3086827
50. Wei, L., Jin, L., Luo, X.: Noise-suppressing neural dynamics for time-dependent constrained nonlinear optimization with applications. IEEE Trans. Syst. Man Cybern. Syst. **52**(10), 6139–6150 (2022). https://doi.org/10.1109/TSMC.2021.3138550
51. Lu, H.Y., Jin, L., Luo, X., Liao, B.L., Guo, D.S., Xiao, L.: RNN for solving perturbed time-varying underdetermined linear system with double bound limits on residual errors and state variables. IEEE Trans. Ind. Inform. **15**(11), 5931–5942 (2019)
52. Li, S., Zhou, M.C., Luo, X.: Modified primal-dual neural networks for motion control of redundant manipulators with dynamic rejection of harmonic noises. IEEE Trans. Neural Netw. Learn. Syst. **29**(10), 4791–4801 (2018)
53. Shi, Y.-H.: Brain storm optimization algorithm. In: Proc. of the 2nd Int. Conf. on Advances in Swarm Intelligence, Chongqing, China, pp. 303–309 (2011)
54. Zhan, Z.-H., Zhang, J., Shi, Y.-H., Lin, H.: A modified brain storm optimization. In: Proc. of the 2012 IEEE Congress on Evolutionary Computation, Brisbane, Australia, pp. 1–8 (2012)

# Chapter 6
# Conclusion and Future Directions

## 6.1 Conclusion

This book is aiming at advancing latent factor analysis for high-dimensional and sparse matrices. In particular, we mainly introduce how to incorporate the principle of particle swarm optimization into latent factor analysis, thereby implementing effective hyper-parameter adaptation.

In particular, it proposes four different hyper-parameter-free latent factor analysis (HFLFA) models. The first model is a *Learning rate-free LFA* ($L^2FA$) model, which introduces the principle of PSO into the learning process by building a swarm of learning rates applied to the same group. This model avoids the choice of learning rate when building a LFA model based on an SGD algorithm. The second model is a *Learning rate and Regularization coefficient-free LFA* (LRLFA) model, which build a swarm by taking the learning rate and regularization coefficient of every single LFA-based model as particles, and then apply particle swarm optimization to make them adaptation according to a pre-defined fitness function. It provides the possibility for different types of parameter adaptation. The third model is a *Generalized and Adaptive LFA* (GALFA) model, which implement self-adaptation of the regularization coefficient and momentum coefficient for excellent practicability via PSO. However, the model hyper-parameters, i.e., $\alpha$ and $\beta$, are remains open since their adaptation strategies are not compatible with that of controllable hyper-parameters. The last model is an *Advanced Learning rate-free LFA* ($AL^2FA$) model. Before building this model, we first propose a novel $P^2SO$ algorithm by incorporate more dynamic information into the particle's evolution of a standard PSO algorithm, thereby avoiding accuracy loss caused by its premature convergence. And then, we adopt $P^2SO$ into the training process of an SGD-based LFA model for making learning rate adaptation with high efficiency.

In general, the goal of our work is to implement efficient hyper-parameter adaptation via PSO for an LFA model built on HiDS data. The readers can

Y. Yuan, X. Luo, *Latent Factor Analysis for High-dimensional and Sparse Matrices*, SpringerBriefs in Computer Science, https://doi.org/10.1007/978-981-19-6703-0_6

immediately conduct extensive researches and experiments on the real applications data involved in this book.

## 6.2  Discussion

There are several research directions that can be conducted in the future.

(a) We plan to develop the potential of other intelligent optimization algorithms, i.e., a beetle antennae search algorithm [1] and a brain storm optimization algorithm [2], to build an LFA model with hyper-parameter adaptation.
(b) We plan to refer the idea of P2SO and incorporate more 'dynamic' information into other improved PSO algorithms [3, 4], thereby achieving further performance gain.
(c) We plan to investigate the ability of PSO in addressing more complex problems such as high-order tensor representation [5–8].

## References

1. Jie, Q., Ping, W., Cpa, B., Chen, G.: Joint application of multi-object beetle antennae search algorithm and BAS-BP fuel cost forecast network on optimal active power dispatch problems. Knowl.-Based Syst. **226**, 107149 (2021)
2. Sun, C., Duan, H., Shi, Y.: Optimal satellite formation reconfiguration based on closed-loop brain storm optimization. Comput. Intell. Mag. IEEE. **8**(4), 39–51 (2013)
3. Memon, M.A., Daula, M., Mekhilef, S., Mubin, M.: Asynchronous Particle Swarm Optimization-Genetic Algorithm (APSO-GA) based selective harmonic elimination in a cascaded H-bridge multilevel inverter. IEEE Trans. Ind. Electron. **69**(2), 1477–1487 (2022)
4. Zhao, D., Cai, C., Li, L.: A binary discrete particle swarm optimization satellite selection algorithm with a queen informant for multi-GNSS continuous positioning. Adv. Space Res. **68**(9), 3521–3530 (2021)
5. Wang, P.P., Li, L., Cheng, G.H.: Low-rank tensor completion with sparse regularization in a transformed domain. Numer. Linear Algebr. Appl. **28**(6), e2387 (2021)
6. Zilli, G.M., Zhu, W.P.: Constrained tensor decomposition-based hybrid beamforming for mmWave massive MIMO-OFDM communication systems. IEEE Trans. Veh. Technol. **70**(6), 5775–5788 (2021)
7. Luo, X., Wu, H., Wang, Z., Wang, J.J., Meng, D.Y.: A novel approach to large-scale dynamically weighted directed network representation. IEEE Trans. Pattern Anal. Mach. Intell. https://doi.org/10.1109/TPAMI.2021.3132503
8. Luo, X., Zhou, M.C., Xia, Y.N., Zhu, Q.S.: An incremental-and-static-combined scheme for matrix-factorization-based collaborative filtering. IEEE Trans. Autom. Sci. Eng. **13**(1), 333–343 (2016)